A Right To Property

Also by Trip Elix

Extortionware: A Hackers Tale

A Right To Property
Its your information,

D.Y
Publishing

This book is dedicated to Aaron Swartz.

You died way too young and were persecuted for your beliefs; you will never be forgotten.

Table of Contents

"Privacy is not dead, people just blindly comply with myths devised to steal their personal information."

Acknowledgments

The forces that are behind the writing of this book came from a group of strangers that I do not know. I simply would state one problem and listen to people for solutions. That is what our politicians are supposed to do; instead they are more interested in putting their hands into the pockets of big companies that steal our information.

The problems that we face as a society are vast, in the finite world of data and information, it is a little easier. Joe Finkle stands out, who wrote a post on a yahoo board as one of the countless members of the Internet community. It saved me hours of time researching alternatives in a subject that I don't use every day. I would like to thank every member of the Internet community. Your efforts and your time is what make the Internet both interesting and engaging.

My close friends, Nelly, Frank, Lori and Patrick who without your support I would have never completed this work. My mother for her wisdom and caring nature into everything I do. Additionally Lisa, who helped edit and Andy Thibault my personal literary mentor.

Introduction

Many think that privacy is over in the United States. Currently, the amount of data collected on every American citizen is astonishing. Government agencies and private corporations together have breached the boundaries of public trust. We live in a world that no longer affords us of our own privacy. Or does it?

I do want to express that I am neither a globalist nor a legal scholar. I proclaim to be a common citizen of this country and the world, one no better or wiser than you. Those who oppose this book and its ideas will proclaim that I am not qualified or educated; I hereby acknowledge that fact and make absolutely no claims to the contrary.

Those responsible for creating systems that track every facet of our lives so completely and have created a society that has become more intrusive than even George Orwell could have imagined.

I understand data, its collection, and the systems that use the data. In the beginning of the digital age before the internet, I worked for one of the first data brokers in the country. My job was to assemble information on individuals, and the company sold the information to private investigators and law enforcement. Since changing occupations, I have watched the processes used in data collection actually become commonplace.

I know very well what systems there are, and how anyone can purchase that information. It was previously my job to do exactly that.

There are things we can do to fix the broken system. Privacy is not a distant whisper in the wind of history. It is a natural state of our beings. It is normal to not want someone to shout our intimate affairs from our rooftops, let alone make each of our lives into a reality TV series.

You have been lied to and the forces in place do not want you to realize the simple truth that you have known all along. You are not powerless; there are things we can do together to strip the power from those who pry into our daily activities.

Throughout the world, there are privacy laws. It exists in legal systems modeled after out own, yet we don't have one.

Much of the information contained in the government should be open and viewable by anyone. Some of the information is private information about us as individuals and is no one's business.

Inside, you will discover that change is needed not only for yourself, but for our government. So much needed data is unavailable, things that should be open for access for anyone but are hidden.

I personally hope you enjoy this material and it will help bring the change we all need.

Property vs Privacy

"Power is not a means; it is an end. One does not establish a dictatorship in order to safeguard a revolution; one makes the revolution in order to establish the dictatorship. The object of persecution is persecution. The object of torture is torture. The object of power is power"
George Orwell 1984

You have the right to be left alone and not to see advertising at every turn. The privacy of your home and every move is being invaded, your private information is being sold to others. Your name is your property. The information that comprises your identity includes your name, date of birth, address, Social Security Number, or any other piece of data that separates you from another person and are the most important details of your existence. Throughout this book, I will make reference to stolen information. It describes data that was taken from you without your consent or knowledge.

Not all information on the Internet is being stolen. Many pieces of information have been given freely by the owners for the entire world to see. That is the beauty of the Internet—it provides a platform that allows mass communication to happen both instantly and globally.

I will make no secret in that I am an Internet fan; I logged on the second day it was available.

There are many different opinions on what exactly the Internet is; it sparks controversy and ignites revolutions. My personal perception of the Internet is that it is the single most deceitful medium in the world, yet capable of uniting the world like nothing else before it ever could.

Your information is for sale on the Internet. Detailed data about you, the ones you love, and everyone else you know are for sale. The same corporations that sell it have no shame, no morals, and make judgments that affect your life. In testimony given to a committee of Congress in 2013, Pam Dixon of the World Privacy Forum stated that companies had sold information that contained the names and addresses for recent rape victims and police officers. Unfortunately while doing my own research, I found that the depths by which the systematic analysis of our information currently exists to predict not only your future spending habits. It can predict many other things in your life, records of habits and soon it will be commonplace for employers and others to purchase these details and use them over the work force.

Part of this prediction is currently in place. Whether you realize it or not, you have multiple credit scores. The credit system as we know it is in peril. Its old mainstay of spying on our every move and selling the rumors and fictitious information now has competition. The so-called pseudo-credit scores exist outside of congressional regulation. Some insurance companies and employers

will use this information in lieu of a financial credit report. These secondary reports are cheaper and include other information including on-line habits, and other things in general you would not expect. The traditional credit reporting systems are becoming increasingly insignificant. Since the general public is unaware of the pseudo-scores, brand awareness will be reinforced by the traditional credit reporting industry. It is the business strategy of the traditional credit reporting bureaus that brand awareness with credit scoring to combat the marketplace dependence on pseudo-scores. The bureaus will adopt advertising to reinforcing brand awareness of credit scoring with consumers. Even though consumer have been kept completely in the dark how the score is tabulated and it will eventually fail.

Another term comes to light when trying to describe how a pseudo-score is derived. You may have heard the term big data; it is a broad term for data sets so large or complex that they are difficult to process using traditional data process applications. These sets of information are combined into smaller sets to allow correlations to be found. The analysis of these data sets allow trends to be discovered, which are open to many kinds of uses. Law enforcement uses big data to predict criminal behaviour. The NSA uses it to predict home grown terrorist threats. Marketers use big data to predict your buying behavior. A simple purchase of baby wipes at a local drugstore by a bachelor yielded ads for other baby items on various search engines. You may have noticed by now that your search results for the same terms will yield different results than other people. The advertising you see on several websites on the internet has been tailored for your consumption.

Your information, or rather the information about you, is out of your control and is available for sale. Some of the companies holding your information may be unknown to you, others may use services or vendors that violate your trust while skirting regulation. These companies acquired your data from years of the banks and the telephone industry selling and publishing your personal information. The major credit reporting companies also sold the names and address history of every client for many years.

You should know who has your information. There should not be a process where it is socially acceptable to spy on your neighbor and sell their intimate details. Today this unacceptable behavior has become commonplace, and it is coming from a very few number of companies that are taking advantage of millions of Americans.

In order to understand who these companies are, it is necessary to break down and compartmentalize what they do. Some companies do more than one task; the industry is large and vast. In some cases, the retail giants that expanded into the credit card system are also participating. To further complicate the complexity of who is doing what, there was the Internet business consolidation of the 1990s.

The companies involved in selling our personal details are many. A few of them operate on the Internet and openly sell information. The industry is aware of the potential backlash from citizens knowing what they are doing. Part of the industry has been given the light of day and exposed in newspaper and television reporting. As a result, many have offered opting out. In this process you must provide your driver's license and other personal details in order to prove that you are who you say you are, and you can opt-out of having your information sold. The problem with this concept is at the time of the writing of this book, there are well over 800 different companies that openly sell information on the Internet. 180 of the companies have an opt-out system. You the consumer would need to opt-out of every single one of them one at a time. The companies do not make it easy; they often will require more information to verify who you are. On the outside, it looks like the opt-out process is actually accumulating data collection and selling your information one more time. The other issue is more obvious; the industry is allowing opt out as a solution because they know they have no right to your information. It is the feeling that opt-out programs will protect the organizations from litigation.

One example that evolved into the mass harvesting of personal information is the direct mail list industry. Have you ever received a piece of junk mail? One of the largest collections of personal data comes from this very industry. Direct mail in the age of the Internet is in its decline. Magazines and mass media have consolidated, and regularly sell its subscriber lists. Trade groups and special interests find a source of income by selling who is in the group. Magazines typically have a narrow interest readership. Inferences can and have been made about the individuals in the lists. The very large media corporations that used to tailor junk mail now offer analytics through the use of big data to analyze your behavior.

One search I found on the internet to me was quite funny. The site was long and provided a vast amount of free gifts just in exchange for the visitors email address.

> FREE Gift #3 ($19.95 value): Without a Trace: The Ultimate Guide to Digital Privacy in the Age of Big Brother.
> Right now, there seems to be no limit on government spying! But the NSA and IRS are not your only concerns…
> The clandestine PRISM program, according to The Washington Post, allows the NSA to access email, documents, audio, video, photographs and other personal information that unsuspecting citizens have entrusted to well-known companies including Apple, Facebook, Google, Microsoft, Yahoo, Skype, YouTube and Paltalk, among others.
> My report helps you easily drop off the digital grid—so no one can infiltrate your privacy and put you at risk!

This site was created by a list broker. All over the internet are sites created by list brokers. The simple virtue of joining and giving your email address, will add your email to a list. Your email address is matched to your name and the rest of your information. Some sites just grab your internet address. Just by opening a site can put you on a list. The list brokers do not even bother to ask your permission. There really are no boundaries for the list brokers, and everyone's information is for sale. One advertiser states that they have information on 292 million Americans, 90 million of whom had recent purchases. The information was available by SKU#. That is the item number that is printed on most items retail packaging.

I have a website set up already that lists many of the known companies. I will be the first to admit that it is not complete. It is difficult to identify every single company involved in the process. The URL is http://itsmyinfo.org. There is a link on its homepage to data brokers; there, I attempt to list every entity that sells our data. If you know of a company not listed, please send me a note, I would be happy to add it.

For a few years, I worked for a company that had bought and sold information derived from credit agencies and other organizations that sell information. I worked as a skip tracer, looking for people who had defaulted on a loan or were wanted by law enforcement. At that time, I realized that the risk of someone stealing another's identity was inevitable. It was before the Internet came to be commonplace. I believe that the threat currently posed by the epidemic of identity theft is a result of open information systems. These systems, and the amount of information they hold, constitute a threat to national security as well. With very little money, anyone can purchase the names and any other details of current members of the military, police departments, and national security agencies; it is all available.

The breadth and depth of information available for sale is almost the complete details of people's lives. The data is not limited to addresses and names of family members. The information captured from police departments' license plate scanning systems are available as well. If you are unfamiliar with license plate scanning, many police departments throughout the country are active in recording the whereabouts of virtually every vehicle. Cameras are positioned at intersections, on poles, and mounted on police cars. The camera records a picture of the vehicle and the license plate. The picture can also identify who is in the car if it is moving, or who is next to the car if it is parked. Computer software then analyzes the picture and picks out the license plate. The data from the plate is then used to search the local jurisdictions department of vehicles. For local police departments, the system allows instant checks on stolen vehicles and enforcement of various motor vehicle laws.

The data from the transaction, the plate, geolocation, and time of day are stored on a third-party computer. The information is for sale. What is worse is that the Department of Motor Vehicles in many states routinely sells all of its information. The information includes the name, type of automobile, license plate number, and other information. The information can be unified with a plate number and give the whereabouts of any individual in their automobile.

Many people in America are upset about police surveillance. These high-tech tools to combat terrorism have been put in place by our government. The revelations brought by Edward Snowden centered on the NSA programs of surveillance of American citizens are distressing. It is not the government that is making a judgment of whether you go to church or engage in risky behavior. These again are used in your pseudo-credit score and are used against you. It is also not the government that is making a profit from selling your information to your insurance company or your prospective employer.

The government's management of our local municipalities and the county as a whole is too large an endeavor to exist without information about its population. The information in many cases is crucial for our dealing with each other. We live in a free society with open records. Thus, even with promises of confidentiality, our information remains prime for the picking. It's been available since the beginning of our nation. However, since the invention of the computer and internet, it's been available for sale. While I strongly advocate for freedom of information, I believe we should reject being exposed, objectified, and monetized using that information. You might believe that there would be some legal framework to protect you. Many have heard of the Right to Privacy Act. It is often cited in the general public. Indeed privacy has been a concern in government for a very long time. There is not a mention of personal privacy in the Constitution. It is often cited that privacy comes from the fourth amendment. What is often overlooked is how our Bill of Rights and our Constitution are applied. People or corporations are covered under our legal system and protected from our government. That is the intention of both the founders of our government and many within our government today. Confusion sets in when one citizen does something to another, like a major corporation harvesting personal information about another citizen.

The Constitution and the Bill of Rights do not apply because neither of the participants is the government. I cannot express how many times I have heard, "those in the know," completely miss this point. I have even watched videos on YouTube from universities that have made classroom seminars for students interested in the legal topics. They will cite the Bill of Rights as if it were between individuals. I have heard the same rhetoric coming from television reporters. Privacy is not a topic that is understood very deeply by those in the media. The media has always been on the forefront of invading the privacy of the population. Newspapers were the originators of much of the concern of privacy in 1880. It was the year that the first photograph was printed in a newspaper.

Government and Privacy

"Relying on the government to protect your privacy is like asking a peeping tom to install your window blinds."
John Perry Barlow Co-Founder EFF

The protection of privacy has a deep heredity within the government. The concern over the privacy of the population was inevitable during the middle of the second Industrial Revolution that occurred over 100 years ago. Judge Thomas McIntyre Cooley wrote in A Treatise on the Law of Torts: Or the Wrongs which Arise Independent of Contract, 2nd edition 1888, that "Personal Immunity the right of one's person is said to be a right of complete immunity: to be let alone." "The Right To Privacy" was originally published in 1890 by Louis Brandeis and Samuel Warren in the Harvard Law Review; it describes privacy and its condition at length. Instead of Congress acting on the suggestions from the writing, it has been left to be cited anytime anyone merely mentions privacy. What was adopted instead was regulation that would be part of the solution to anyone peering into our most private of affairs.

Instead of regulation becoming the solution, Congress has redefined credit reporting and allowed it to become the most nosy and intrusive organizations set upon the same society, it set out to protect.

The regulatory efforts of Congress imposed upon the credit and banking industries happen so frequently that we have permanent committees within both the House and Senate. The abuse of the general consumer is a constant struggle that government efforts contend with and attempt to balance. There should never be a consideration made by the government pitting privacy of the population against economic decisions. The primary function of the credit system determining the likelihood of repayment for goods has been lost. It has instead crept its way into the cost of goods. It also decides who can work or be educated. Without public assistance for college loans made by the government, it would be virtually impossible for anyone in lower incomes to receive higher education. Part of the dependency on government finance has to do with the restrictive policies of credit reporting.

One of the best legal descriptions of privacy, as it applies to the general population coming from a government prospective, comes from an interview broadcast on CSPAN. The current Supreme Court Justice Antonin Scalia who, while describing the changing direction of the court, stated "conversations are covered by this vague right of privacy that is contained in the Constitution." His choosing of the word vague comes from his years as an educator in the legal field, and a sitting justice of the highest court of the United States.

The recording of conversations was referenced from the court's ruling on the interpretation of the Fourth Amendment of the Constitution of the United States. The Amendments to the Constitution are referred to as the Bill of Rights. At the time of the writing of those documents the general concern that was of an oppressive government not unyielding interests of corporations. The Bill of Rights is a list of limits on government power. The majority of the Constitution, including the Bill of Rights, dictates rules for the people of the United States in its dealings with the entity called the government, not with each other. This concept is difficult for many people to understand or grasp the afforded rights of the population are often misrepresented by those that educate. Take, for example, the foundation of our privacy on how we interact with the government:

> The right of the people to be secure in their persons,
> houses, papers, and effects, against unreasonable
> searches and seizures, shall not be violated, and
> no warrants shall issue, but upon probable cause,
> supported by oath or affirmation, and particularly
> describing the place to be searched, and the persons
> or things to be seized.
> --Amendment IV U.S. Constitution.

The Fourth Amendment only protects you against searches that violate your reasonable expectation of privacy. A reasonable expectation of privacy exists if you actually expect privacy, and your expectation is one that society as a whole would think is legitimate.

The expectation of privacy ruling came as a decision of the United States Supreme Court in 1967, <u>Katz v. United States</u>, holding that when a person enters a telephone booth and shuts the door to make a call, the government cannot record what that person says on the phone without a warrant.

In this case, the recording device was stuck to the outside of the glass in a phone booth and did not physically invade Katz's private space. The Supreme Court decided that when Katz shut the phone booth's door, he justifiably expected that no one would hear his conversation and that it was this expectation, rather than the inside of the phone booth itself, that was protected from government intrusion by the Fourth Amendment.

This idea is generally phrased as "the Fourth Amendment protects people, not places."

You may expose a lot more than you really know or intend to about yourself to the government. Most information that a third party collects—such as your insurance records, credit records, bank records, travel records, library records, phone records, and even the records your grocery store keeps when you use your loyalty card to get discounts—was given freely to them by you, and is probably not protected by the Fourth Amendment under current law. There may be privacy statutes that protect against the sharing of information about you at state level. Some communication records receive special legal protection, but there is likely no constitutional protection. It is often very easy for the government to get a hold of these third-party records without your ever being notified.

The Fourth Amendment clearly is not meant for protecting information held by third parties. The misunderstanding and misrepresentation of law and its purpose is rampant within American society.

Prior to 1974, many computer systems had been installed in the various departments of the executive branch of the federal government. The information free flow that followed included the IRS, and it was common place for information to be disclosed to people who didn't have a right to the information. The passage of the Right to Privacy Act solved some of the federal data leakage. It also instituted a general misconception within the population. Many people will refer to this act as if it somehow protects them. The act limited the disclosure of information by the executive branch of the federal government only. Additionally the Privacy Act officially dehumanized the people of the United States by lumping all of us into a computerized system. This act contains three major categories; "system of records," "statistical record" or "routine use."

The problem with the Privacy Act is that the American people believed it in a broader sense. It did achieve privacy and a right from the executive branch from disclosing countless details like census and IRS records. There is disassociation of what the United States government has institutionalized as privacy. The average U.S. citizen's sense of privacy and belief and expectation of it are not in concert with reality.

We live in a land under one of the most unique systems of government in the world. Life, liberty and the pursuit of happiness were not merely mentioned in the Declaration of Independence in passing. "They are endowed by their creator with certain unalienable rights." This was the signed document that set this nation apart from others. Now ask yourself fellow citizen what is your most important possession?

Perhaps with careful examination of your own inventory you may overlook your own name. It is the basis for what comprises your identity. Everyday however our identities are literally up for sale, which undermines the well-being of each and every citizen as well as our economy and American society itself. This threat has existed for decades and is on the verge of becoming a national tragedy.

We are afforded rights and privacy is not among them. It is not mentioned in any of the writings. The Constitution only mentions property once. It is wrongly assumed that the Founders did not consider property rights important. Actually property rights were address in many ways in both the Constitution and Bill of Rights.

The importance of property is recognized in every branch of our federal government. The courts focus on property at almost every turn. Yet privacy is not mentioned often, and your right to your own name which is the key to your own existence within the courts is ignored.

The laws of the United States are derived from what is called common law. Common law is the basis our very legal system. Definitions and concepts exist in common law that impacts us all. The idea of what is understood as right or correct within with society comes from many different influences. Not all of these are written rules within our legal system.

Under a free society it is not the natural order of government to only to prescribe to what has been written, as a singular guide to existence. Lawmakers and the courts switched the actual intent of the framers and created law to be conceived in a literal sense, it is also known as doctrine of absurdity. Forever changing the intention of the framers that rights had to be expressed to be granted. This ideal is slowly creeping in society into our higher court systems.

The Video Privacy Protection Act of 1988 was passed in reaction to the disclosure of Supreme Court nominee Robert Bork's video rental records in a newspaper. The Seventh Circuit Court of Appeals dismissed a class action against Redbox in 2012, and ruled that the Video Privacy Protect Act was "not well drafted." The law was written to protect the records of people that rent video tapes. The Seventh Circuit found that the Video Privacy Protection Act does not allow customers seek damages when video-rental businesses keep rental histories longer than allowed.

A three-judge panel said the law allows damages only when video services illegally disclose viewing histories and other personal information, but not when they violate a requirement that they destroy the records a year after use. The decision overturned an Illinois federal court ruling in the case.

The Act is not often raised, but stands as one of the strongest protections of consumer privacy against a specific form of data collection. It prevents disclosure of personally identifiable rental records of "prerecorded video cassette tapes or similar audio visual material."

Privacy has always been an expected state of the basis of our own beings. It is an inherited ideal which in my opinion the framers took for granted and did not feel the need to express. There is an idea that we all possess of what private means to us individually. You know in your person what private means to you.

Another area of protection through regulation occurs with medical records. The HIPPA (HEALTH INSURANCE PORTABILITY AND ACCOUNTABILITY ACT OF 1996 Public Law 104-191 104th Congress) and the later the Affordable Care Act (Patient Protection and Affordable Care Act H.R. 3590 111th Congress, 2009–2010) regulated medical records. The laws were created to regulate and protect patient confidentiality.

As a society, we are led to believe in the confidentiality of our medical records during and after our office visit. Often insurance policies require a co-payment at the doctor's office. Paying with a credit or debit card may expose the doctor's name and what they practice to the credit industry.

Drug stores and other merchants carry loyalty cards that house and openly market consumer data that has shown to be a pervasive problem that sidesteps the protection of regulation. A simple application at one of these stores to obtain a few cents off of a dollar fuels several multi-billion dollar industries that sell your information without your consent. Credit reporting companies and credit merchants also collect and sell the information about any individual purchasing any type of medication.

Practically all loyalty card programs, sell the data they collect.

It is not only the drug stores and loyalty card programs that give up customer information. A number of websites have been set up to support people with various ailments. Some of these were put up by drug manufacturers. Some of these were created by the data brokers to simply collect information. What is gathered however, are lists, there are lists of people available-for-sale for almost any ailment. Alcoholism, cancer, mental illness and many more, exist in lists outside of the protection of HIPPA. This was also testified to in congress in 2013. I personally believe that Congress should make a sweeping change to the legislation and include everything that we use on or for our bodies. Toenail clippers, hairspray, and toothbrushes should fall under HIPPA protection. It is no one's business outside of our doctor that we use or consume any of these or any other medical items.

Lists categorize you as an individual and with whom you reside. Credit reporting companies have free range to assign new residences to your credit score whether or not you lived there. That's just a residence; what about other inaccurate information that may cause serious disruptions in your life or those of your family that are listed in public documents? A delinquent bill such as a telephone bill may be inaccurately reported to the credit bureau in a minor's name when it was actually the legal responsibility of the parent. Having public information so readily available to these credit companies can create a wide array of mishaps that are not regulated. Regulation does not limit the collection of data only the dissemination of it.

The value that the credit system places on society is misleading. It makes selling to those who have bad credit a threat to merchants. This logic has been a significant bone of contention since its inception. We are forced to fuel the system when we enter the marketplace. We unknowingly become the commodity in our endeavors for daily goods. Simple coupons, discounts, or warranties appear attractive until we realize how far and in how many hands our information travels.

As an innocent consumer, we authorize corporations—without knowing who their clients, contractors, or business partners are—to record our information. Behind the scenes, business intelligence on clients equates to big money. Depending on the company, it is not necessarily the product that the consumer is buying that is important to meet the bottom line. In actuality, your consumer information can be bought, sold, or traded to the point where the value of what you purchased has nothing to do with you walking into the big chain conglomerate. In the electronics industry; you may have seen a product for sale less than what you can buy it for on the Internet. It is important for some of the industry to identify who the buyer is. Your short-range purchase was staged for a longer-term gain. People who fall for the trap are categorized as one more data notch, one more personality trait.

The problem with privacy is that there is no such mechanism for compensation by the actions of another within that form of framework. Property, however, has a long history of compensation for the use of it.

The very idea of your information as a property right is not new. The United States Senate has maintained ownership of this idea for over 30 years. On February 8th, 1971, The Wall Street Journal printed a story entitled Sen Ervin vs. 'Information Power' by David C. Anderson. It stated that "one privacy expert suggests that a person's right of privacy of information about himself ought to be given the legal status of a property right"; this was testified before Congress the day before the article appeared.

Property assignment began with the name you are known by. There is nothing written in the law to guard your identity otherwise. The arbitrary set of information is used to identify one individual from another is known as identity. Our current system is set up so that you need to utilize your identity for the very depiction of yourself to others.

Property and its ownership are the basis of the framework of the Constitution. It was the general belief that delegation of powers under the Constitution protected individual property and the rights of individuals. The concept of contracts has its basis in property. Property rights are paramount within our society. The arguments within our nation and its policies are made balancing the rights of individuals and property. The founding fathers considered the right of property as a basic human right.

James Madison was one of the signers and is considered the father of the Constitution. He first was a member of the House of Representatives and, in later life, our fourth United States President. While he was a House member, he wrote the first ten amendments to the Constitution to form the Bill Of Rights. He did this because of tension between the federalists and the anti-federalists who held that a Bill Of Rights was necessary to safeguard individual liberty. The Federalists argued that the Constitution did not need a Bill of Rights because the people and the states kept any powers, not given to the federal government. He wrote this passage in Federalist paper 62 when speaking of the formation of the Senate:

> The internal effects of a mutable policy are still more calamitous. It poisons the blessing of liberty itself. It will be of little avail to the people, that the laws are made by men of their own choice, if the laws be so voluminous that they cannot be read, or so incoherent that they cannot be understood; if they be repealed or revised before they are promulgated, or undergo such incessant changes that no man, who knows what the law is to-day, can guess what it will be to-morrow. Law is defined to be a rule of action; but how can that be a rule, which is little known, and less fixed?
>
> Another effect of public instability is the unreasonable advantage it gives to the sagacious, the enterprising, and the moneyed few over the industrious and uniformed mass of the people. Every new regulation concerning commerce or revenue, or in any way affecting the value of the different species of property, presents a new harvest to those who watch the change, and can trace its consequences; a harvest, reared not

by themselves, but by the toils and cares of the great body of their fellow-citizens. This is a state of things in which it may be said with some truth that laws are made for the few, not for the many.

In another point of view, great injury results from an unstable government. The want of confidence in the public councils damps every useful undertaking, the success and profit of which may depend on a continuance of existing arrangements. What prudent merchant will hazard his fortunes in any new branch of commerce when he knows not but that his plans may be rendered unlawful before they can be executed? What farmer or manufacturer will lay himself out for the encouragement given to any particular cultivation or establishment, when he can have no assurance that his preparatory labors and advances will not render him a victim to an inconstant government? In a word, no great improvement or laudable enterprise can go forward which requires the auspices of a steady system of national policy.

It is obvious by this passage that former President Madison did not anticipate the cultivation of society, being the product by which the government had allowed to be harvested. It is equally as obvious that the power of corporations was not thought of as a competition to governmental power.

The illusion made from the framework of our very legal system is broken. It is defies simple logic that individuals can create lawsuits, own property, and enter into contracts and what they are called by as individuals or the description of them is not upheld as their own legal property.

Through the duration of time, our individualistic role in society created an informational exchange based on a "fee for goods" society. As individuals, we have become the commodity, and there are high financial stakes. We have inherently fallen victim to an informational era where data is created from our own homes. All of this without any form of compensation.

3rd Party Privacy

"Credit is an 'I love debt' score."
Dave Ramsey Radio personality

There are regulations that were designed and implemented to guard personal types of information. The various credit reform acts limit access to consumer credit reporting information by third parties. The total lack and disregard of securing credit card transactions has fostered a worldwide epidemic in stolen credit card numbers. Large industries have often influenced legislators to pass criminal laws to protect the interest in the information it gathered. It also sways efforts in the legal systems to protect its interest by criminalizing activity.

The banking industry is no exception, and the monetary system of the United States is tied heavily to credit. The United States Secret Service and our tax dollars are used to enforce the banking system's own problems. It is far cheaper to lobby Congress and have tax dollar enforcement than to pursue action civilly or fix its own security issues. The protections given to the credit industry by the American government are unprecedented. The United States government invests heavily in the protection of money and has employed several counterfeit counter measures over the years. All forms of payment outside of money are vulnerable to simple forgery and with modern computer technology, it doesn't take a rocket scientist to figure out how to be a criminal. The banking industry has spent virtually nothing to protect its monetary instruments. To further add insult to injury, any costs associated with any new implementation are passed directly to the merchant. Junkies and meth heads commonly employ vulnerabilities in the banking system that only end up costing the merchant and the taxpayer money. Checks are easy to make fraudulently, credit companies regularly send cards to addresses other than the one listed on the individuals credit report just to name two problems there are many of them.

The government protects the monetary system and considers it a matter of national security. It has been the long-term goal of the credit and reporting industries to tie itself to the banking industry and to be institutionalized to the degree that it is also, protected by the federal government.

Individual safety and security are eroded any time the government gives preferential treatment toward outside institutions over the protection of the people. At times, the general misconception surrounding the credit system is that it was created so that the majority of laws protected the individual. There is not anything in U.S. law that protects the ownership of a person's name or data identity or the rights thereof.

For most the adherence to credit scoring is critical. The credit system does not care about you. Your life, family and friends, have become only data points to a system that treats us all like an oversized ant farm. You must use cards you have in your possession. Failure to adhere to credit card usage rules will influence the cost of car insurance. Listen to Clark Howard on the radio who rails against this practice weekly.

The influence of big finance and the credit reporting system is not out of the courts either. The very idea that an individuals identity does not have any value comes from the belief that your name is public information. Public information was defined by the Freedom of Information Act at the federal level. The act was instituted in 1972 under President Johnson. Before the act was passed, the definition of public information was the same as marketing is today. Companies would have divisions that were called public information, and those departments had public information officers.

Multiple court rulings set the precedence surrounding the protection of the credit reporting industry. As a result, all three credit reporting agencies lost an unfathomable amount of consumer data. The overall wisdom of court rulings provided only a couple of years of credit monitoring service for the millions affected. We have been so influenced by the credit system that our worth as humans has been defined by our credit scores.

The value placed on our credit scores has placed an undue burden on those with lower income. Because of the value, many in lower income have a higher threat of identity theft because the popular belief is that they are limited to the amount of credit they can get. Identity thieves are well aware of this and take advantage more often than higher income groups.

The rise of identity theft and compromised credit information has inadvertently allowed the reporting companies to influence Congress. The established credit reporting companies are under regulation, however the alternate is considered self-regulatory. Under the established regulated industry, it is difficult at best to rectify incorrect and/or inconsistent credit information. The alternate credit reporting industry has no checks or balances; the consumer doesn't even know who to contact. The information is hidden; my personal phone call to one company denied that they had my records. I was able to talk to someone who eventually advised me otherwise.

The very perception of credit is that it should be private and our own property. As a forced participant in modern day society, we hold a misconception that credit is available to us at our own free will. Using the Internet, it is dictatorship control that prevents constant timely access to our own information. There is no excuse in the Internet age that anyone should not be able to check their credit record any time they want and as many times as they want for free. You are able to see your own information more than once a year if you pay for it. Each and every one of the former lawmakers should be ashamed of themselves for caving into the greed of the regulated credit reporting systems.

One of the most confusing factors for the consumer marketplace is that there are multiple credit scores holding different values. The credit reporting system is a very profitable industry. Merchants, insurance, and employers pay to gain credit reporting information. The consumer becomes a dual-faceted horse that further runs the machine at one end; the merchants feed the information to the system and reporting comes out the other.

Under the Fair and Accurate Credit Transactions Act (abbreviated FACT Act or FACTA , Pub.L. 108–159), an amendment to the Fair a Credit Reporting Act (abbreviated FCRA) passed in 2003, consumers are able to receive one free credit report per year from each of the reporting agencies. Previously, consumers had difficulties in obtaining their own information and had to pay for their reports. In the beginning, the credit reporting agencies offered dysfunctional web interfaces and sleight of hand tricks. After the passage of FCRA, internet websites promoted free credit reports in banner advertisements on major search engines and landing pages. Experian offered a "free credit report," which would enroll users into a credit-monitoring service. In 2005, Experian accepted a settlement with the Federal Trade Commission that it had violated a previous settlement. As a result, it fostered another industry.

The alternative credit monitoring services on the internet bypass the permission of the consumer and almost all the protection of the regulatory system. As a result, the consumer information was offered up for the sole discretion and ownership of the data collector. A search of "free credit report" in any internet search engine will reveal many of these companies. Many of the companies on the internet are out-right scams and are in place to steal your information.

Others are in place to circumvent regulation and shunt your credit information from the regulated side to the unregulated side. Often these services are paid quite handsomely. How else could you explain the budgets in advertising on television for a service that is free on the Internet?

The companies that exist in the credit reporting system are also data brokers. Virtually anything is available from them outside of credit. For many years the credit reporting system sold individual information outside of credit reports and for a wide array of purposes.

The FCRA limits the use of the credit report to certain purposes. They are:

- Applications for credit, insurance, and rentals for personal, family or household purposes.
- Employment, which includes hiring, promotion, reassignment or retention.
- Court orders, including grand jury subpoenas.
- "Legitimate" business needs in transactions initiated by the consumer for personal, family, or household purposes.

- Account review. Periodically, banks and other companies review credit files to determine whether they wish to retain the individual as a customer.
- Licensing (professional).
- Child support payment determinations.
- Law enforcement access: Government agencies with authority to investigate terrorism and counterintelligence have secret access to credit reports.
- Specific prior consent is required before consumer reports with medical information can be released.

All three sell credit header information. A credit header is identifying information from a credit report. It includes name, mother's maiden name, date of birth, sex, address, prior addresses, telephone number, and the Social Security Number.

Target marketing is not a permissible use of credit reports. Currently, both Equifax and Experian are in a consent agreement with the FTC to not use credit reports for target marketing. Trans Union attempted to challenge the FTC prohibition on using credit information for target marketing but failed in Trans Union v. FTC.

All three are in the list business and actively sell mailing lists for virtually anything else outside of credit reporting information. I personally have seen mailing lists being advertised for sale for who has a disease by type of ailment. The sale of lists for spam and junk mail are a mainstay of profit for the industry which will custom create any list the purchaser desires.

In 2015 the FTC did a follow-up to its 2012 Study on

Credit Report Accuracy. It reported on its website:

> The 2012 study found, among other things, that one in five consumers had an error that was corrected by a credit reporting agency (CRA) after it was disputed on at least one of their three credit reports. The study also found that about 20 percent of consumers who identified errors on one of their three major credit reports experienced an increase in their credit score that resulted in a decrease in their credit risk tier, making them more likely to be offered a lower auto loan interest rate.
>
> The follow-up study announced today focuses on 121 consumers who had at least one unresolved dispute from the 2012 study and participated in a follow-up survey. It finds that 37 of the consumers (31 percent) stated that they now accepted the original disputed information on their reports as correct. However, 84 of these consumers (nearly 70 percent) continue to believe that at least some of the disputed information is inaccurate. Of those 84 consumers, 38 of them (45 percent) said they plan to continue their dispute, and 42 (50 percent) plan to abandon their dispute, while four consumers are undecided

Some of the sites on the internet that sell information about individuals get the information directly from the credit reporting companies. Identity theft itself became institutionalized along with the selling of personal information.

If you have ever become a victim of identity theft and tried to get your valuable documents back that were designed to prove your existence, you then fall prey to being a victim of big government as well. For the first time, you realize you no longer have ownership of your own identity, and you need to be the bearer of proof. (You will understand that your very existence is within the system of paper and data is met without merit towards yourself.) No matter who knows or cares about you, your left alone in the virtual data desert.

Once your social security number is compromised, you are stuck with it, and the circumstances surrounding it can last a lifetime. As a member of society, you as an individual become indebted to a system that holds the number assigned to you as your personal debt regardless of who created it. There is a general assumption that the very government that assigned our number will provide a sort of protection of the very number (social security number) that should be secured and protected. It was commonplace in a number of states to use that number in many ways including voting registration. It is now assigned at birth, and there is no changing it regardless of how it has been compromised. The millions who have fallen prey to identity theft know this all too well.

Perhaps in a century someone will realize that the social security numbering system needs to be replaced. I would highly recommend it be a new number that is given along with a cipher. Doing this would ensure that the original number would only be used with the government and the cipher could be regenerated at will.

Hopefully, you never have to attempt to change your number with the always friendly and overly bureaucratic Social Security Administration Office. You will find out how easy getting a new number is after waiting for hours to get denied. There is not a real fix for identity theft within society or the system that collects our information. Those that collect information don't differentiate between thieves on your account or your actions. Very simply put, they don't care about you; it's your actions they are after.

We unknowingly became a functional commodity of the proverbial machine. Multibillion-dollar industries have sold our data. It erodes our practical perception of our beings leaking into our thoughts and dreams. It turns us all into a monetary stream, hidden by the guise of a nonexistent definition of public information. It is celebrated by the industry and falsely linked to the economic dependence.

In an age where an individual's actions are cataloged and traded as a profitable exercise, is it not damaging to the individual to not be compensated for their very labor they provide by existing and being active in that very same system?

It is an insidious idea that individuals have the right to own property, but their name is not a form of it. Many would think incorrectly of legal fiction that governs corporations or Intangible property of copyrights or trademarks. Tangible personal property refers to property, except land or buildings that can be seen, weighed, measured, felt, touched, or otherwise perceived by the senses. Individual names should be considered as tangible non-transferable personal property as well.

The United Nations lists a basic human right. You have the right to own things, and nobody has the right to take these from you without a good reason. It should also be noted that the protection of intellectual property also falls within international human rights law.

The concept of your name as your property is not difficult to grasp in government either. Most of us have visited a public park. The very idea of someone harvesting the lawn at a park and selling sod is the same craziness that we as individuals experience with our data and the government.

You do own your name it is your property. If we can convince congress to change it to law, the impact would be enormous. It would take away the careless attitude about storing other person's property. I have heard many times that companies do not care about security simply because they know that no civil action would take place if they lost another's property. Imagine how many sites that have lost personal information would react. If those that lost data had the worry that each person's life that have been potentially devastated. It would allow those that are affected to have a form of recourse. It would also begin the process of limiting the amount of information exposed.

The responsibility for safeguarding information exists with each and every company that holds our collective information. The loss of information that surrounds the news reports is evidence that companies do not care about your information that they may lose. The threat of fines by the government or civil litigation seem to be the only moral limits that many companies use to guide their actions.

It would be the very reason that both political parties would not embrace this idea. Both receive large contributions from these very companies. The politicians also embrace using the information for their election campaigns.

Those that gather information without permission could have civil remedies emplaced for stealing. Many in the merchant card services industry do that now. Outside of the applications installed on smart phones and computer browsing tracking, the merchant card services collect and sell our location and buying habits. A merchant card service is the institution that accepts credit cards for the merchant that you buy thing from.

In the data world, there are a number of things about you that describe you; these things are called the keys to your identity. There is a distinction that should be made; in the data world, there are a total of two other entity types that you interact with, the government and others. Your information, the stuff about you, belongs to you. You as a member of society share the use of your information with the government. All others have no right to your property unless you allow them to use your information. Many privacy and confidentiality agreements state openly that they share information. Sharing information is not the same as selling information. Many of the companies that exist with loyalty card programs market identifiable information to a wide array of manufacturers of products. The information is sold under the guise of market intelligence. Our agreement with the merchant in most cases has been breached. Some merchants are finding that selling product with zero profit actually is more lucrative because they collect the information from the buyer of the product and can resell that information over and over. Our data is being captured. In the data world, there are many systems that have no right to use your informational property.

There are data quarries that capture your identity and are housed in data warehousing systems which are for public and private auction. Information surrounding your name is cherry picked. The majority of the information in reality is technically stolen. There are other pieces of information that are used to identify you or single you out from other individuals. There are other identifiers such as your mailing address, email account, or cell phone number that are used by a wide array of systems that marks or flags your identity.

The government holds proprietary ownership of your social security number, and it allocated the address at which you reside. Your birth certificate states your name and date of birth and is virtually the title to your name. Any and all other information acted by you is your own information; it does not belong to anyone else.

In the world of data, this is not a question of whose information you have. Those that sell our information should be on notice that we the people, are not for sale.

A Not So Free and Open Government

"He has called together Legislative Bodies at Places unusual, uncomfortable, and distant from the Depository of their public Records, for the sole Purpose of fatiguing them into Compliance with his Measures." Declaration of Independence 1776

Any individual should be able to inspect the information housed in all levels of the government. If the information disclosed is in the course of profit, then all of those who are listed in the records should be duly compensated. The only exception should be those that use the information in the course of business that do not directly resell it. This very statement is intended to send chills down the backs of all of those who resell information about the population.

Our government is in the data business. Bureaucracy depends on collecting information from the population for its daily work. At the local level, there are all kinds of records, including property records, arrest records, and tax records. Many of these records are needed in the course of other businesses.

The United States Government is the biggest hoarder of general information in the world. One agency the NSA, is in the process of building the world's largest data center. The Colorado River was redirected to cool the massive computer center. It is estimated that it uses 1.7 million gallons of water a day to cool the center. It is one the few man-made objects that are recognizable from space. This level of hording data is unparalleled globally.

There are many records that are necessary for a free and open society to thrive. The information that the government collects should no longer be restricted. Public information should be available to the public. Reporters and others inspecting the records held by agencies of the government have trouble finding proof of governmental waste and or corruption. A free and open government requires that all records with few exceptions are open for inspection. The very idea of inspecting individual records is not the same as the mass harvesting of our population. At the federal level, the Privacy Act protects information that is held within agencies of the executive branch. The privacy act does not cover your city or town tax office. Government employees should recognize the difference between simple observation and the mass harvesting of information.

The information we share with the government about ourselves belongs to us. It does not belong to the public or the government. We inadvertently give the information and do not expect it to be used against us as citizens in our country. Congress should deny anyone who attempts to collect information from the people in bulk and recognize it for what it is. It is a massive stealing of property from the people of the United States.

Most of the state governments are freewheeling with the data it collects about its population. The Department of Motor Vehicles often will not disclose singular information about an individual over privacy concerns. The stalking incidences that resulted in murder influenced several states legislators to enact anti-stalking legislation in the 1980's and 90's. It blocks access to license and motor vehicle data in the given state; however, the state will sell the entire collection of information including names, addresses, vehicle ID number, color, and license number to data brokers without any such concern. Voting records and political affiliations of individuals used to be regularly bought by politicians when running for office. The higher offices afford larger budgets and individual targeted advertising was applied to several million people in 2010 during the presidential election. These individuals had tailored advertising created for their consumption.

The quantity of information that the government collects is only one side of the equation. It also contracts with others to collect information. Some companies eagerly create low-cost systems for the use of the government, and the information can be harvested as part of the agreement.

There are multiple systems that are paid for by the taxpayer. The grant system is known throughout academia. Grants make it possible for researchers in physics, medicine, robotics and numerous other fields to do their work. More often than not individuals in these fields have been required to enter their findings into research systems.

It is all too often that the same researchers cannot profit from their own work. When other individuals want to use the research, it has to be paid for. It is a sad state of affairs when hard working Americans strive for a higher level of education and the research that the individual conducted is not accessible to the general public or to the author.

The university libraries have multiple ways to store journals and papers from higher education. Many of the systems offer so-called centralized storage. The distant database's where information is compiled restricts the information or papers to the physical location of the library where it was placed. Students and researchers find this out and realize that the work is locked away at a distant university or perhaps does not exist any longer. One of these systems is called JSTOR.

JSTOR started as an organization with offices in New York City and Ann Arbor, Michigan. It was originally funded by the Andrew W. Mellon Foundation and began in 1995. It was housed at seven different libraries that included topics such as economics and historic journals. In 2009, JSTOR merged with the non-profit Ithaka Harbors, Inc.

For many years, anyone in higher learning has been forced to depend on the collection and citation from JSTOR or one like it. Pieces of significant achievement are generally made through higher learning. Thankfully, there were not databases in the time of Galileo or Da Vinci.

Academia is a driving force to further society. It is unfortunate that countless parts of the much-needed information was written and placed where only the select few are allowed access. The very fact that individuals pay taxes should make one believe that even our general population should be allowed to gain accessibility to this material. If you are not currently a student or know someone who is, then the research that was done is not available. JSTOR was not only strongly marketed throughout college campuses, but it became a staple in the creation of college papers.

JSTOR and systems like it allowed libraries to outsource the storage of various journals, publications, and associated research with the confidence that these documents would remain available for the long term. The JSTOR website had a statement that the enclosed content was provided by more than 900 publishers. The database contains more than 1900 journal titles in more than 50 disciplines. In 2012, it created a program allowing for the viewing without copying or printing of three older articles in its collection for new users to the system.

Citizens started to become frustrated at the government and the entanglements of corporate interest. The Occupy Wall Street group is just one example of frustration. There continues to be movements that hold the belief that corporations are regularly allowed to commit piracy in the real sense of free data. Private and public companies have gathered and housed disclosable information and turned it into a profit center. The public domain becomes held hostage in the process. Corporations regularly use the judicial system to enforce claims on various data that is publicly disclosable or is paid for by our tax dollars.

There is a cultural misunderstanding within society that the internet was created by big business. It is fostered by the huge conglomerates that act as toll barriers to the open system. Tim Berners-Lee invented the World Wide Web. It is a hypermedia initiative for global information sharing. He wrote it in 1989 while he was at CERN, which is the European Particle Physics Laboratory. He was the author of the first web client and server in 1990. His specifications of URLs, HTTP, and HTML were refined as Web technology spread. His efforts were donated to the world for a better humanity.

Aaron Swartz has been referred to as an internet pioneer and prodigy. He worked on various projects that became the basis for website functionality. At twelve years old he founded info.org. It was a website that encouraged people to post information they knew about. He hoped that others would want that information. It was created several years before Wikipedia was founded. He won a prize through a school competition sponsored by the web design firm ArsDigita. He took his first trip to Cambridge that would later prove to be the city of his undoing.

Many of the platforms and things that people see on the internet are created by individuals that work in groups. These groups are open to anyone that wants to contribute. During the development of RSS and XML, Aaron acted as a member of one such community. He worked on the project remotely with the team of people who were mostly based in California. The team was primarily a group of accomplished programmers who were in their late twenties and thirties and were astonished to find that he was only thirteen.

Later he worked on the development of Creative Commons, a type of licensing agreement that is used by millions of individuals on the internet. Traditional copyright rules and corporate conglomerates still attempt to enforce old rules and regulations around the guise of sharing information. The Creative Commons License allows the author to decide how his or her work can be used by others.

Swartz had studied ways to work with large data sets. In 2008, his interest was in national politics and founded watchdog.net. The site stated that it was, "the good government site with teeth," to aggregate and visualize data about politicians including where their money came from. In most writing about Swartz, this website is rarely mentioned. It is likely that the political repercussions from this site contributed heavily to the inevitability he had later in life.

He also worked with Shireen Barday at Stanford Law School to assist in the research in "Problems With Remunerated Research." It appears in the Stanford Law Review Volume 61, Issue 3. He downloaded and analyzed over 400 thousand law review articles that purportedly were taken from Westlaw's collection. Swartz uncovered troubling connections between funders of legal research and favorable results within the courts. The document is available to read on the internet.

Swartz had placed a laptop with an external hard drive in a basement storage room at the Massachusetts Institute of Technology (MIT) campus. It was plugged into an unprotected network switch. The laptop was copying large amounts of data from JSTOR. On January 6th, 2011, he was arrested by the MIT police. The state dropped the case against Swartz when a federal prosecutor became interested. Swartz was charged with the maximum penalty possible. The charges carried penalties of fifty years of imprisonment and one million dollars in fines.

Swartz founded Demand Progress that was created to win progressive policy changes for ordinary individuals through organizing and lobbying. It focused on issues of civil rights, liberties, and governmental reform. The campaign was created to lobby against the internet censorship bills SOPA and PIPA.

The Stop Online Piracy Act (SOPA) was a United States House Bill introduced by U.S. Representative Lamar S. Smith (R-TX) to expand the ability of U.S. law enforcement to combat online copyright infringement and online trafficking in counterfeit goods.

The PROTECT IP Act (Preventing Real Online Threats to Economic Creativity and Theft of Intellectual Property Act, or PIPA) was a United State Senate Bill introduced by Senator Patrick Leahy (D-VT) and 11 bipartisan co-sponsors. It was a proposed law with the stated goal of giving the US government and copyright holders additional tools to curb access to "rogue websites dedicated to the sale of infringing or counterfeit goods", especially those registered outside the U.S.

January 18th, 2012 was a monumental time in history because the internet operators and grassroots organizations caused politicians to change their direction. Wikipedia denied access to its own service. The big and small players included Google, Tumblr, and Twitter as well as others that advertised banners on the home pages against the passage of SOPA and PIPA.

Lobbyists held a strong-hold on both political parties. The SOPA and PIPA bills had substantial support and were expected to pass. Protests over the legislation occurred not only in New York but San Francisco, Seattle, and Washington D.C.. The House and the Senate finally lamented to pressure and withdrew.

Aaron Swartz was a media fixture during the protests. His appearances on many of the daytime talk shows made his name known to millions. His criminal case was led by a federal prosecutor named Stephen Heymann. The final offer was provided by the prosecutor's office and was reduced to one felony charge. Aaron would commit suicide, two days later on January 11th, 2013. He strongly believed in public access. He took his own life due to the bullying process that has been easily allowed by the United States Government. The motives are unclear to this day. His legacy will remain unfulfilled due to corporate greed.

Several individuals have criticized that the charges were overzealous. It was displayed on the sound bites that were given as homage by all forms of media the day of his death. A White House website petitioned to fire Mr.

Heymann due to the way the case was handled. The site generated more than 25 thousand signatures in less than a month. He is still in office as of this writing.

Aaron Swartz took an alternative approach to changing the problem of public and free access as the author of the Guerrilla Manifesto. It has been suggested that the justice department became worried over it spreading into a movement. You can find the writing on the internet.

Swartz gave a speech at Illinois University on October 16, 2010. He stated:

> "By virtue of being students at a major U.S. university, I assume that you have access to a wide variety of scholarly journals. Pretty much, every major university in the United States pays for these sort of licensing fees to organizations like JSTOR and Thompson Isi to get access to scholarly journals that the rest of the world can't read."

> "These licensing fees are so substantial that people who are studying in India, instead of studying in the United States don't have this kind of access. They are locked out from all of these journals. They are locked out from our entire scientific legacy. I mean a lot of these journal articles, they go back to the Enlightenment. Every time someone has written down a scientific paper, it's been scanned, digitized, and put in these collections. That is a legacy that has been brought to us by the history of people doing interesting work, the history of scientists. It's a legacy that should belong to us a commons, as a people but instead, it's been locked up and put online by a handful of for-profit corporations who then try

to get the maximum profit they can out of it."

The multiple university library systems that house work is important to us. Over the years we have switched over to digital copies only. This too is a foolish idea. There should be systems of long-term storage in alternate mediums. A simple cataclysmic event could potentially erase our shared legacy. Fortunately, the systems are not housed in the same geocentric location. The over dependence in electronic and magnetic storage is a reality. Physicists and Astrophysics both agree that the likelihood of our sun projecting an electromagnetic pulse (EMP) is a matter of time. The effects of such an event are deeply speculated. Some hypnotize that the entire electrical grid globally would be affected. The hubris behind supporting systems that could simply be erased by nature is only a position that could be undertaken by institutions of higher education.

Courts and Criminal Data

"So the public domain should be free to all. But it is often locked up. You know, there is often guard cages. It's like having a national park with a moat around it and gun torrents pointed out, in case somebody might want actually to come and enjoy the public domain."
-- Brewster Kahle founder of the Internet Archive.

The identity of criminals is something that most societies have always made available to its population. In modern society, this very form of information can be difficult to find even though it has been recorded by the government.

During the computerization of government records in the 1960's, the FBI instituted the National Criminal Information Center. The following description is taken from the FBI website.

> "In the 1960s, Director J. Edgar Hoover presided over the meeting during which the decision was made to implement a computer system that would centralize crime information from every state and provide that information to law enforcement throughout the nation. Working with the International Association of Chiefs of Police, the FBI created an advisory board made up of state and local police to develop nationwide standards and consulted with

the Commerce Department to build an effective telecommunications system. On January 27, 1967, the system was launched on 15 state and city computers that were tied into the FBI's central computer in Washington, D.C.—which at that time contained five files and 356,784 records on things like stolen autos, stolen license plates, stolen/missing guns, and wanted persons/fugitives. In its first year of operation, NCIC processed approximately 2.4 million transactions, an average of 5,479 transactions daily."

The center was created as a closed system. Only those in law enforcement or involved in security can access the records. Individuals can view their own records by making a request to the FBI. However, information that is held may be omitted at the discretion of the agency.

The process of data entry at the beginning of computerization had errors in the records. Any time humans do any form of data entry manually, errors are created. It is common for individuals to be mis-identified or be hooked into other information that was not theirs. The early regulatory efforts of the credit systems were brought about in part for that very reason. The creation of the National Center of Criminal information also had problems of mixed identity within its system. It also has no way to detect that the information is wrong.

The public deserves access to information about criminals that are among us. Employers, educators, and individuals that are vulnerable are just a few of the examples of who needs to gain access to criminal records.

It is estimated that 50 to 55 percent of criminals inside the United States have records in the NCIC system. Those that are experienced in performing background checks for the government and the military are encouraged not to depend solely on the NCIC for information and are encouraged to use other systems as well.

There is not a national database that contains all of the criminal records in the United States. The data that exists in many systems is faulty. It is not even designed to locate Information that is associated with a conviction. A legal process can be overturned or expunged. Felony convictions that are listed can be downgraded to misdemeanors based on the judges' discretion. The systems have many problems.

Several websites and others in the field have incorrectly identified individuals and potential criminal associations with absolutely no oversight. There is not a way of checking if an identity is mixed up with someone else's. It becomes troublesome when the naïve owner of their own information is represented of having a criminal presence.

The aftermath leaves someone that is unemployed and helpless as a victim once again where records are captured and housed on various databases. There is not an easy way to fix this issue. Three credit conglomerates provide methods of correction once noticed. Prospective employers simply set the applicant aside to not go any further. Individuals looking for work are left clueless as why they were not considered for employment since they are innocent and a victim of not their own doing.

Individuals who commit crimes that involve sex are considered social lepers. Many states have adopted the sexual registry and made the information available on the Internet. Individuals may be required to report where they live and work. On the surface of this, it all would sound reasonable. The bad and evil sex convict should have his or her own records available. Those that prey on children should be watched, and the public should know where they live.

The sexual registry has several problems. One of the largest issues is that not everyone that is in every registry is a sexual criminal. Some prosecutors use the registry statistics in politics. It is often advertised how many predators they incarcerate at election time because society views sex crimes as being of the utmost importance. Some prosecutors' cheat the statistics by convincing individuals with low mental capacities or no real representation that have committed a crime to plea to a sex crime not fully understanding what stigma will be placed on them. There is a substantial issue at hand in the world of data. The Internet holds data that merges without discretion or oversight with other information. Addresses of residence, typos on apartment numbers, certain parts of a phone number can and have gotten individuals who don't know each other merged in the databases.

The problem with criminal records is they require much more work than is deemed necessary. The NCIC should be open and corrected so that individuals can see their own information. The United States is a big place, with many more systems of government than most people realize.

There are over sixty territorial and state systems. There are thousands of local courts in addition to the federal court systems. All of these individual systems of the government emanate information that is crucial for society to function. Criminal information is housed at many of these.

The earliest records that exist of mankind were created from the population and its form of government. Scrolls have been found in ancient China and Egypt. Unbelievably, five thousand years later, it is still debated that the government held records and information paid for by the public should not be freely disclosed.

The courts are also a source of criminal recourse and a host of other information. The federal court system utilizes PACER. The description from its website states:

Public Access to Court Electronic Records (PACER) is an electronic public access service that allows users to obtain case and docket information on-line from federal appellate, district, and bankruptcy courts, and the PACER Case Locator. PACER is provided by the Federal Judiciary in keeping with its commitment to providing public access to court information via a centralized service.

PACER has had issues since its inception in 1988. It was a system that was accessible by various terminals in libraries and office buildings. In 2001, PACER was available on the Internet. It has been plagued with problems. The ten cents a page fee includes a horrible search method in addition to documents that are not correctly encoded, and it is slow. In August of 2014, it finally received an update and then deleted landmark decisions from its database in the process.

Portable Document Format (PDF) has been used for decades as the default way we look at documents. A scanned or digitized image of a page of text is actually a picture. To convert it to text, Optical Character Recognition (OCR) is used. Pictures files can have stamps, wrinkles and folds. Evidence of stamps and other features of scanned images should be maintained. The format of the file doesn't allow tagging text within a picture and then storing the computerized text along with the picture in the file. A cross-platform hypertext document storage file format doesn't currently exist for pictures of documents.

OCR has difficulty deciphering characters that have lines or folds in them. One solution that the court has never implemented is to specify a single font to be used on all documents submitted. The system has many files stored within it that are pictures of documents. To circumvent the deficiency in the format itself, metadata is implemented.

One of the complaints about PACER is that many of the documents are stored as picture files with incorrect document encoding. PACER is not a research system. It houses over 500 million documents that contain every motion, brief, filing, and ruling of every case in the federal courts, 99% of which is useless to the public at large. PACER is actually a dumping system and not intended to be used for legal research, which is unfortunate.

Due to the deficiencies and lack of planning within all levels of government these problems exist. The federal court system is only part of the problem; PACER is the federal court database. States, cities, and towns all have information that needs to be collected indexed and housed. There are companies that collect information from those sources also. They then claim copyright and ownership on the collection of information. There are other ways to do legal research. Westlaw and Lexis are big alternatives to PACER and both sell access to the information. Incredibly some attorneys attempt to claim copyright on legal briefs they file with various courts and attempt to file suits against the pair. The selling of court documents is lucrative; both companies make millions of dollars a year. The profits on court information made by various companies and the PACER system itself each year surpass ten billion dollars.

The very idea that systems need to exist to correct and append the collection of court records is perverse. PACER is not broke; it operated a yearly surplus of over 150 million dollars, before its last increase in fees in 2011. It is one of the few government run services that actually create a profit every year.

There are tools that are used by federal attorneys every day. These tools are offered as a pay-for-system. Most do not offer any free equivalents. The process of finding out if a statute, case, or federal regulation is valid may require the use of third party tools. Shepard's Index is owned and published by LexisNexis and KeyCite that is published by Westlaw are the recognized authorities in the field. LexisNexis and Westlaw search interfaces provide useful proprietary tools that do not compare to free or cheap alternatives. Both use vast resources to compile and organize informational sources from all over the nation. The organizations capture cases published by local individual courts, property registration with various county clerks all over the country, and many of other sources of data.

To protect the information companies require signed agreements not to distribute the data they have copied. You can duplicate this legwork and gather documents online from many State Court systems or go to the Courts and look at hard copies for older cases. This is time-consuming and not practical in most instances.

Some of the information is captured by agreements that companies have with the court. The transcription service, for example, is paid for by the court. The vendor who then resells to the data to a provider who makes a profit selling access to the records the court already paid for. Some of the systems are purchased or leased by communities to manage cities, and towns are owned by companies who then sell the information.

One alternative for doing research is Loislaw. It offers a flat monthly fee for doing research. Researching the federal court information is problematic. Advocates for free access to court records created a program called the PACER recycling project its acronym is RECAP (PACER spelled backward). The project features plug into Chrome and Firefox web browsers where users could upload documents they had already paid for.

In 2009, Carl Malamud, a technologist, author, and public domain advocate, joined forces with Aaron Swartz and others on the RECAP database. Carl Malamud helped put EDGAR, the Electronic Data Gathering, Analysis, and Retrieval system operated by SEC online. He was a visiting professor at the MIT Media Laboratory and is the former chairman of the Internet Software Consortium. He was instrumental in creating the first Internet radio station along with many other achievements.

The YouTube movie "The Internet's Own Boy: The Story of Aaron Swartz" features a quote:

"PACER is just so awful" ... "The system is 15 to 20 years out of date, "said Malamud. He was quoted in a story that was written by John Schwartz on February 12th, 2009. It appeared in The New York Times. The Title of the article was "An Effort to Upgrade a Court Archive System to Free and Easy." The companion photo that ran with the article was of Aaron Swartz.

Swartz had downloaded close to 20 million pages of text before the FBI contacted him. Swartz was not charged, but the FBI investigation ended Swartz's involvement. After the incident, the FAQ section on PACER's homepage warns that unauthorized attempts to collect data may result in criminal prosecution.

All of court information is public domain. It is amazing at best to state otherwise. The unprecedented greed of the United States Senator's own creation is appalling.

In May of 2014, Princeton University's Center for Information Technology Policy and the non-profit Free Law Project announced a partnership to operate RECAP. The Free Law Project took the lead on developing and maintaining the RECAP system. Steve Schultze was one of the original creators of RECAP. He stated, "Harlan, Tim, and I are delighted that RECAP will be stewarded by the excellent team at Free Law Project, which shares our audacious vision of making the public law free to the public."

Cornell is the best free alternative to LexisNexis and Westlaw for statutes and commentary. It contains annotated statutes for most States, a free version of the bluebook citation guide, and many major public domain publications available from the ABA and other sources.

The issues with access to court information are so vast that even Google decided to create a solution. Google Scholar is not as complete as LexisNexis or Westlaw; however, over time and based on popularity, it will improve.

All of these systems are needed for society to function. Our court records are a complete mess and strangely enough generate billions of dollars in profits each year. All of it is befitting a lawyer joke.

Private companies not only sell information but collect it. The systems exist inside of offices of public servants throughout the country. The contracting company then turns local clerks into publicly paid data entry personnel. This is done for property record transfers, tax records, and a wide variety of information gathered at the local level.

Throughout the country, Homeland Security has been giving grants to local communities for collecting license plate data. It is called LPR, which stands for license plate reading. The system uses computers and cameras to read license plates automatically. Cameras are installed on local police vehicles and at many intersections. The cameras photograph cars and their license plates very quickly, some report the capabilities of up to a thousand plates per second. The system installed in the police car communicates with the databases of the police department and motor vehicles along with the vendor to look for infractions or other crimes. The system then will inform the police officer if the vehicle is stolen or didn't comply with other laws like taxation or registration.

The vehicle then can be stopped by the police officer. Some municipalities are using this technology for parking enforcement. The parking compliance officer merely needs to drive past cars in a given area and knows instantly if they have paid for parking.

The system records the plate number and geo location (longitude and latitude) of the police car. Some also record the direction of travel of the license plate. The system cameras and computer software many times are purchased or are leased by government from private companies.

The information is often shared with the company providing the service to the police department. The cost of storage and the systems themselves are enormously expensive.

By using a shared computer software system, police departments can afford to install and use the information to solve crimes. It has often been reported in newspapers and interviews by police officials justifying using those systems.

The ACLU in many areas has filed suits against police departments for storing license plate information too long. To solve this issue the private companies sometimes just limit the amount of time that the police can look up records that are housed. In most cases, the police paid by taxpayers are collecting information for an outside company to then profit from. The ACLU has filed suit against the police in the city of Los Angeles. The police are tracking every cars movement in Los Angeles County calling it an active investigation. The police are reading approximately three million license plates a week. The case will be going to the Supreme Court of California as of this writing.

The information as it is collected by the course of police business is actually creating a profit for the companies that house the information. Some of the agreements that police departments sign include provisions of ownership of the information and termination provisions.

Some of the companies involved with this industry also sell the information to other companies outside of law enforcement. Repo operators have used the information to locate and repossess cars that have been marked as in default by the banking system. Although the tracking of vehicles is in its infancy, the total effect of this practice will only be shown as a positive effect on society. One of the reasons for this is that the agreement between companies and municipalities bar the mere mention of the name of the company. The general agreement given by one company has this strange covet. Vigilant Solutions has one such disclosure addendum in their agreement.

5. PROHIBITIONS OF DISCLOSURE BY USER

5.1. You shall not create, publish, distribute, or permit any written, electronically transmitted or other form of publicity material that makes reference to LEARN-NVLS or this Agreement without first submitting the material to LEARN-NVLS and receiving written consent from LEARN-NVLS. **This prohibition is specifically intended to prohibit users from cooperating with any media outlet to bring attention to LEARN-NVLS.** If you breach this provision LEARN-NVLS may terminate this Agreement immediately upon notice to you.

Vigilant Solutions was started in 2005; because it is a private company, it does not have any oversight by congress. It is not subject to any regulation. Records used by official police investigations are obtained without a simple chain of custody protocol being followed. The company is not an official member of any law enforcement bureau or an official member of any court. Attorneys may be interested in using their data in actions involving motor vehicle incidents. Members of congress and high-level bureaucrats should be aware that their movements just might be for sale.

Those interested in purchasing the movements of others go to the web site owned by Trans Union. http://tlo.com The going rate for a vehicle history is a mere ten dollars. There are too many examples of mislaid public money contributing to private companies to list.

Private companies and corporations are intertwined with our government at every level. Many use the public's resources in for-profit systems, and then use our justice system, investigators, and courts against us to safeguard the threat of competition or their own existence all the while using our tax money.

The idea of open information and the way our government treats so-called public information is abysmal at best. Information that is captured by the government should be disclosed by the government. In this state of technology, it is not only possible to make this happen, but there are many experienced veterans of programming wanting and ready to help. We the people should have free access to all of it. No individual concern should be harvesting our information to market it. If we were considered the owner of the information much of the harvesting would disappear. Your name is not public information.

The United States Government has created systems that have taken the nation to the moon. Perhaps that lesson of building the necessary components at NASA should be a lesson for our legislators. You don't have to be a rocket scientist to see and understand that government data has problems.

The Journey to Surrendering Privacy

"The cost of freedom is always high, but Americans have always paid it. And one path we shall never choose, and that is the path of surrender, or submission."
John F Kennedy

The first bank loans date back to 2,000 B.C.. Thus, for almost 4,000 years, lenders have collected a variety of personal data on borrowers.[1] This practice applies to other transactions too, from basic service providers (measurements for tailoring clothes) to retail transactions (address, license number, social security numbers, even references, etc.) and of course, almost anything to do with government agencies.

Across that history, our information has been embedded within different realms throughout the credit system. The practice of creating a credit score is attributed to Lewis Tappan. He was said to have hated credit. However, he realized that offering credit to customers was becoming the only way to make a sale.[2] In a similar vein, the conception of a formalized credit bureau was formed by a group of London tailors in 1803 who compiled information on customers who failed to pay their bills.³ The Manchester Guardian Society, formed in 1826, would become part of the foundation of Experian, one of the leading American credit-rating agencies more than 150 years later.⁴

In the US, during the 1850s, the first credit agencies emerged, with the aim of transforming a broad variety of information into a product that could be sold for profit. One of the earliest US credit agencies, Merchants' Credit Association, was founded in 1897 by Jim Chilton. Nearly a century later, this would become another fundamental part of Experian.⁴

There has always been a deep-seated fear of government and spies within our society. In 1949, Eric Arthur Blair, better known as George Orwell, published his novel, 1984. Its dystopian vision was a rallying cry against the institutionalization of bureaucracy. Fear of "big brother" remains a recurring concern in society, though punctuated by long periods of acceptance of the status quo of bureaucratization that sometimes seems to exceed Orwell's literary forewarning. Americans seem to waver between the promised benefits and the downright caustic intrusion of surveillance measures posed by commercial and government agencies.

The inception of The Retail Credit Company occurred in 1899 from the minds of two brothers, Cator and Guy Woolford, who settled in Atlanta, Georgia. They kept a list of customers and their credit-worthiness for the local Retail Grocer's Association. Even before the turn of the last century, they would sell the book to other merchants in the association. Over the years, the Retail Credit Company purchased smaller credit reporting agencies and expanded its business to include selling reports to insurers and employers. [5,6]

The massive size and task of tax collection in the United States required the assistance and categorizing of computerization for the creation of the Internal Revenue System. These records were comprised of governmental documents that were compiled and computerized. The system was installed in 1961 by ADP (Automatic Data Processing)[7], which went public the same year. Millions of tax records suddenly were housed and categorized on punch cards. This system of record keeping would later prevail to be not only time-consuming and inaccessible, but also obsolete. A humorous movie about the system is in the national archive and available on the internet. The title to search for is "Right on the Button."

Many of America's original credit reporting agencies were also debt collection companies. There were hundreds of credit reporting agencies. [8,9] Retail Credit sold stock to the public for the first time in 1965. During this time, Retail Credit also took its first steps toward automation[10] that converted files written on 3x5 index cards into electronic data systems. [11]

By 1966, the abuse of the credit reporting industry was the talk of Congress. Testimony was given to The House of Representatives on consumer affairs of the committee on banking and currency records. Among the findings were that the Credit Data Company had collected records in five cities which consisted of an excess of 9.6 million individual records that were collected in Southern California. In 1967, TRW bought interest in Credit Data with an option to buy the remaining shares.[12]

TRW originated in 1901 with the Cleveland Cap Screw Company, which was founded by David Kurtz and four other Cleveland residents.[13] Their initial products were bolts with heads electrically welded to the shafts. In 1904, a welder named Charles E. Thompson was able to expand upon their process of the craftsmanship of automobile engine valves. In 1915, the company was the largest valve producer in America.[14] Charles Thompson became a general manager of Thompson Products in 1926.[15] The experimental hollow sodium-cooling valves aided Charles Lindbergh's solo flight from New York to Paris across the Atlantic.

In 1950, Simon Ramo and Dean Wooldridge worked for Hughes Aircraft.[16] They spearheaded the development of the Falcon Radar-Guided Missile, among various other projects. The pair grew frustrated with Howard Hughes and his form of management. Howard Hughes was an aeronautics pioneer and directly oversaw many of the projects his companies created first hand. The Ramo-Wooldridge Corporation was created after the two left Hughes with the financial support of Thompson Products.[17] The detonation of a thermonuclear bomb by the Soviet Union spurred Trevor Gardner to form the Teapot Committee in October of 1953. Ramo and Wooldridge were members. Their corporation became the lead contractor of the Intercontinental ballistic missile (ICBM) development effort. The ICBM development was under the direct supervision and guidance of the United States Air Force.

The continued backing from Thompson Products allowed the Ramo-Wooldridge Corporation to develop new solutions which consisted of computers and electronic components. Pacific Semiconductors in 1954 were funded by this corporation.[18] As a result; the production of space crafts were manufactured such as Pioneer 1. Thompson Products and Ramo-Wooldridge merged in October of 1958 to form TRW, which held a half a billion dollars in government contracts when it bought interest in Credit Data.[19] The landscape of the country had a perception of only one large credit reporting company. In actuality, Retail Credit Company operated along with hundreds of smaller credit and collection companies throughout the United States. The Retail Credit Company dominated the market, and the corporation was called a monopoly by members in Congress. Various credit companies inflicted abuse on that of the general public. As a direct result, Congress held hearings and called several credit reporting companies to testify.

The members of The House of Representatives on Consumer Affairs Committee on Banking and Currency commenced hearings on June 20th and 21st in 1968. Testimonies were given that consisted of credit abuse practices being inflicted on the population by various credit reporting bureaus.[20]   Prior to 1968, credit companies were not regulated or operated with a bias toward themselves or each other. Decisions were created in regards to the worthiness of credit. Merit-based credit was based on things such as an individual's sexuality. Many women were barred from their own ability to obtain credit. The system used sexual, racial, and religious markers along with other discriminatory demographics. Many things outside of payment history were considered.

During this time period, the groundwork for being denied credit was commonplace. Individuals were barred from credit based on marital status, use of alcohol or drugs, and many other factors. Parents were affected by the actions of their children.

Before the internet and the era of mass communication, the magazine industry compiled lists of consumer-specific data that housed names and addresses by areas of interest. The consumer believed in a reasonable expectation of privacy and the elongated process of filling out simple forms appeared to be harmless. The conditioning of the American public to fill out questionnaires had begun. The process before computers was slow. It would take months after a form was entered for any mailing to take place.

One of the most notable hearings was held on May 16th, 1968. The president of the Retail Credit Company, W. Lee Burge, testified under oath that it would not be economically feasible for his company's 45 million records to be entered into a computer system.[21]

The Truth in Lending Act was originally Title I of the Consumer Credit Protection Act, Pub.L. 90–321, 82 Stat. 146 was enacted June 29, 1968. The act was crucial because it created first-time access for millions of individuals to gain access to his or her own credit. The act did not cover abusive collection tactics used in gathering and placing data that would later corrupt credit scores.

The Union Tank Car Company was a railcar leasing company that created TransUnion as its' parent company. TransUnion stated on its website:

> "In 1969, we recognized an opportunity to grow a new business using our technical expertise. We acquired the Credit Bureau of Cook County (CBCC), which manually maintained three point six million card files in four hundred seven-drawer cabinets. Soon after the acquisition of CBCC, we became the first company in the credit reporting industry to replace accounts receivable data with automated tape-to-disc transfer, drastically cutting the time and cost to update consumer files. Early in our history, we recognized the enormous benefit that a national, online information system would bring to clients and responded with the first online information storage and retrieval data processing system. This system provided credit grantors across the country with one source for fast and valuable consumer credit information."[22]

Charles Ward and his brother Stephen founded Demographics Inc. in 1969 to meet the needs of the Young Leadership Council for the Democratic Party.[23] The members created political mailing lists that were competitive with a data processing system. A system was devised to compete with what Winthrop Rockefeller and the Arkansas Republican Party had already been utilizing.[24]

The Fair Credit Reporting Act of 1970 was created as a repair of the Consumer Protection Act. The passage of the act and its aftermath allowed the banks and credit reporting agencies to collect information from the public for use of the credit system. There never has been a system in place to protect what information was collected. It was a vagrant disregard to our individual privacy and personal ownership of our name and familiar identities.

More stories of abuse of information surfaced after the passage of the act. Rumors and speculation in some cases led to out-right lying in consumer credit reports.[25] The Retail Credit Company finally entered its paper card files into a computer system that it solely ran and operated in 1970. It only took two years to complete adding the data to a computer system, even though it testified to Congress that it would not be feasible in 1968.[26]

The industry was plagued with abuse. The poor and minorities were not yet able to create a credit score. Women were denied because they were not married. There were inaccuracies that the system refused to fix and in some cases lied about.[27]

By 1972, Demographics Inc. was one of the largest mailing list companies in the country.[28] It was not alone since there were hundreds of competing mailing list companies already in existence.

The Fair Credit Billing Act (FCBA) was enacted in 1974 as an amendment to the Truth in Lending Act. Its purpose was designed to protect consumers from unfair billing practices and to remedy the injustices that occurred.[29] The FCBA was designed to aid women and forbids discrimination and denial of credit based on sex or marital status. It was monumental because it also required credit companies to acknowledge complaints, math errors, unauthorized charges, etc. [30]

In the UK, the business we now know as Experian developed during the 1970's through Great Universal Stores (GUS), an established retail conglomerate with millions of customers paying for goods on credit. Sir John Peace, a computer programmer at the time, helped to combine the mail order data from various GUS businesses and created a central database in Nottingham which later added Electoral roll data as well as county court judgments. In 1980, the decision was made to commercialize GUS' database under the name, Commercial Credit Nottingham (CCN) and the business was led by Sir John Peace.[4]

The founder of Demographics Inc. was in financial difficulty when his main line businesses, Ward School Bus Manufacturing, experienced failure in 1975. Ward divested himself of Demographics Inc. In that same year, Charles D. Morgan was a company manager for three years and became the new president and CEO.

Morgan traveled to the Yellow Pages maker Direct Media Inc. in New York. There he met with David Florence to discuss changes in the operation. A decision was made to focus exclusively on direct-mail technology. Morgan returned with an idea that became the List Order Fulfillment System (LOFS). Direct Media was the sole customer. LOFS had been described as the nation's "first fully-automated, on-line system used to generate mailing lists." By 1978, the company touted that it had created the first comprehensive "marketing database" in the United States. [31,32]

Harte-Hanks Newspapers was founded by publishers Houston Harte and Bernard Hanks in 1924. The firm bought Texas newspapers that included papers in Corpus Christi (1928), Wichita Falls (1948), and San Antonio (1960). A San Antonio TV station was created in 1962. The firm went public in 1972, which was the year Harte died. His son, Houston H. Harte, became the chairman.[33]

Over the next decade, the company diversified into free coupon books and brochures that expanded beyond Texas. The paper began to be sold in San Antonio in 1973. It later changed its name to Harte-Hanks Communications in 1977. In 1984, five executives made the company private again and added 700 million dollars in debt in the process. In 1986, it had 70 newspapers. Harte-Hanks began consolidating operations to reduce its debt and focused on larger markets. By 1988, it had sold off half its holdings and bought others in California, Dallas, and Boston. [34]

By 1980, the company's holdings included 29 daily newspapers, 68 weekly newspapers, four VHF television stations, 11 radio stations, four cable television systems, and three trade publications. Its fastest-growing division was its Consumer Distribution Marketing (CDM) unit. It consisted of its advertising shoppers, three market research firms, and three direct mail distributors. Also included in CDM was electronic publishing and video entertainment software businesses. Moving into the recession of the early 1980s, Harte-Hanks's emphasis fell more heavily on CDM. As one analyst told the New York Times, "Harte-Hanks is much less newspaper-oriented than the other newspaper companies. It's more interested in information transfer. It's much more financially oriented." Robert G. Marbut confirmed this, telling the New York Times, "Our job is not just to produce newspapers. Our job is to meet people's needs for information." [35,36,37] Robert G. Marbut was the then CEO of Harte-Hanks.

Retail Credit expanded into other industries and changed its name on the New York Stock Exchange to Equifax, becoming the largest provider of credit and other services worldwide in 1975. Many believe that change of name was to lose the tarnished reputation of Retail Credit. [38]

In the decade of the 1980's, Demographics Inc. went through two name changes and began selling stock to the public in 1983. It finally ended up becoming the Acxiom Corporation. The creation of this corporation held historical value to the development of present terms used in the marketing world. The terms included data warehousing, consumer intelligence, the word demographics, target demographic research and execution, cross marketing and referencing of client specific information and products sold. Information gathering by various companies became poignant at this time due to such a lucrative industry with a multitude of financial backers.

The 1990's brought upon society the introduction of the internet. Acxiom acquired full ownership of InfoBase. Direct Media Inc. gained the accumulation of consumer data that consisted of names, addresses and phone numbers and Pro CD, Inc.[39] Acxiom/Direct Media, Inc. offers direct email service.[40] It proclaimed publicly on its press release:

> "Acxiom delivers a wide spectrum of data products, data integration services, mailing list processing services, data warehousing, and decision support services to major corporations internationally. With the transaction, Pro CD will have greater resources and consequently, will be able to introduce even more innovative desktop marketing and reference tools on the Internet and on CD-ROM. The Pro CD acquisition reinforces the business strategy and opens up new opportunities for both parties. Acxiom's immediate plans are to enhance Pro CDs data with additional content that adds value to the end user's applications." [41]

Following a change of ownership, TRW was renamed Experian in 1996 and in the same year the business was acquired by GUS and combined with the CCN Group. This deal completed on 14 November 1996. Two businesses, each leading the field in their countries, were brought together under the leadership of Sir John Peace; indeed it was his vision of building the first global information business that has been the driving force behind the creation of modern-day Experian. The big three now dominate the credit industry—Equifax, Experian, and Trans Union.

The internet boom of the 1990's brought in hundreds of those that resell data and really too many to list, though most of them are smaller operations. The turning point came as a result of September 11, 2001. The National Security Agency called it "the program," the mass collection of data on the American population. Internet companies were contacted by the NSA to install backdoors to the systems. The government had instructed internet and telecommunication companies to not only give access but how to create data sets from the systems. The systems were programmed to automatically created metadata summaries of larger data sets. The government had trained private industry the most efficient way to spy on its own customers. By the summer of 2002, The NSA connections to internet and telecommunication providers was complete.[42] Because the relationship the private companies have with the NSA all of the work is classified. The companies, however, found that the information in its metadata format was valuable and stated selling transaction information.

Consumers were left unaware that websites were systematically spying on their activities. By 2008, a new web browser tracking method called a cookie was introduced. A web cookie is a file that is stored on a computer or device to keep track of a user's action on a web site. A cookie is used for a variety of reasons like helping a website remember settings or login information. They also can be used to remember which pages you visited. Cookies do not carry viruses. They are not programs, they are simple files that web browsers use. Third party cookies, however, were concerns starting in 2011 when Congress took only a decade to introduce a failed bill called the "Do Not Track Me Online Act of 2011;" there were a number of failed pieces of legislation calling for the end of tracking web browsing information.

There is another kind of Cookie that is important called the authentication cookie these are used by browsers that use the information in so-called secure sessions. Banks and other institutions depend on the presence of the cookie to know whether or not the same user is logged in or not. These also help the browser keep track of the authenticated cookie using encryption.

Hackers can use authentication cookies to bypass user logins on sites without real security. In 2014, Google identified a vulnerability attack called POODLE, which stands for Padding Oracle On Downloaded Legacy Encryption. POODLE, Google said in a blog post, is a "widespread" threat because it enables prowling hackers to steal your most private information.[43]

There were numerous factors that bullied society into believing that personal information is not under the control of the individual. Consumers discovered that they were powerless. It is more expensive to not comply with many of the systems. The credit system was not alone. Those on the inside of that system were convinced by other forces that people's identities were somehow open for the taking.

The primary force came with the inception of the telephone. For much of its history,[44] The Bell System functioned as a legally sanctioned and regulated monopoly.[45] It was formulated by Bell president Theodore Vail in 1907. It was the telephone that by the nature of its technology would operate most efficiently as a monopoly providing universal service. Vail wrote in that year's Bell's Annual Report that government regulation, "provided it is independent, intelligent, considerate, thorough and just," ... "was an appropriate and acceptable substitute for the competitive marketplace."[46]

One legacy left by the phone company, however, is still present in today's society. The phone book itself is the best example of how a single corporation can legally extort information from the public. Being listed in the directory was mandatory, and the only way out of publication was to pay to be unlisted. The first telephone directory consisted of a single piece of cardboard and was issued on February 21st in 1878. It listed fifty businesses in New Haven Connecticut that had telephones.[47,48] In 1991, the U.S. Supreme Court ruled in Feist v. Rural that telephone companies do not have a copyright on telephone listings because copyright protects creativity and not the mere labor of collecting existing information. The customers all were forced to disclose their information; the court ignored whose property the information was, and the individual did not agree individually to be sold into the abyss.

The act of forcing the publishing of everyone's name, address, and phone number reinforced the demise of the phone book itself. The cellular industry doesn't have a phone book. The threat of regulation has kept the industry from creating the second revenue stream[49] of the industry. The phone carriers also sold the unlisted directory that housed information of individuals that chose and paid to not be listed. I saw one of these personally back when I was in the information business. It looked like a regular phone directory; however, it only housed unlisted numbers that were not made available to the general public.

In the 21st century, printed telephone directories are increasingly criticized as waste. In 2012, after some North American cities passed laws banning the distribution of telephone books, an industry group sued and obtained a court ruling permitting the distribution to continue.[50] The manufacture and distribution of telephone directories produces over 1.4 million metric tons of greenhouse gasses and consumes over 600 thousand tons of paper annually.[51]

The old system of the telephone networks may have competition from cable companies for telephone service but the old practice of extorting customers still exists. It would be insidious to say that the lawyers that are in public service and in every facet of state and local governments are unaware of the definition of extortion. It is good to ponder, how many public officials that have law degrees pay for an unlisted number at their own homes.

The Freedom of Information Act (FOIA), 5 U.S.C. § 552 provides full and/or partial disclosure of previously unreleased information and documents that are controlled by the various agencies and offices of the United States Government. The act defines agency records subject to disclosure; it outlines mandatory procedures and grants nine exemptions to the statute.[52,53] It was originally signed into law by President Lyndon B. Johnson despite his misgivings,[54,55] on July 4th, 1966, and went into effect the following year.[56]

The act fostered a new concept in American history. Information housed by the federal government belongs to the public. It also convinced multiple generations that a new term existed which was deemed public Information. It's heading Title 5 U.S. Code § 552 - Public information; agency rules, opinions, orders, records, and proceedings.

The term *public information* does not appear in the encyclopedia throughout the 18th, 19th and 20th centuries. Many organizations including the military at the time, had public information officers or departments that shaped the perception of its organization to the public. The term marketing came to replace public information within society around the same time as the act was passed. There is a subtitle irony in that those that harvested our information relied heavily on the existence of its own title of yesteryear.

The information housed by the government should actually be called public disclosable information. Public domain objects should not be confused with public information and your name. The idea that our names and identity belong to anyone else is an illusion and should be ignored.

The FCRA[57] (Fair Credit Reporting Act) limits the use of the credit report to certain purposes.[58] The agencies collect much more information and are able to sell that information for a wide variety of uses. The act itself limits the credit information for direct uses. Marketing and the collection and dissemination of information carry no such restrictions.

Banks sold creditor and depositor information to the general mailing industry and many other parties on a regular basis. Regulation was passed barring the sale of some of our information.[59] The Gramm-Leach-Bliley Act's[60] (GLB) added notice and opted out of the provisions that were associated with the Fair Credit Reporting Act (FCRA). The FCRA currently requires that financial institutions make clear and conspicuous disclosures to consumers regarding the sharing of certain information.

The GLB Act requires certain disclosures to be made as part of any privacy policy given to consumers or customers. The industry can change any policy at any time. There are no rules of how the communication is handled. Typically, institutions mail consumers a multi-page document that requires a magnifying glass to read.

The social revolution started ten years ago and is feeding a new avenue other than general gossip that is associated towards your credit file. Many banks are now using social media to gage interest or credit scores.[61] Credit companies have been monitoring social networks for consumer behavior.[62,63] There are a number of banks using social media information to inform on its credit decisions and to determine aspects of its relationship with customers that includes pricing. [64,65]

In 2011, the democratic administration announced that consumers have the right to control which companies collect and use our information. It also stated that the privacy policy of companies should be accessible, transparent, and understandable. It dictated that hacking and personal information leakage should be completely stopped.[66,67]

The Consumer Privacy Bill of Rights was designed to deter internet companies from the collection of personal information for targeted ads. In response, Internet companies such as Mozilla, Google, Microsoft, Yahoo!, and AOL promised to provide a "do not track" mechanism so that customers can choose whether they want to participate in online behavioural advertising or not.[68,69] However, the guideline has the limitation that it is not enforceable. The Obama Administration encouraged the United States Congress to grant the Federal Trade Commission the authority to enforce each element of the statutory Consumer Privacy Bill of Rights. Once enacted, Internet companies infringing upon the rights put forth in these guidelines could suffer sanctions from the FTC.

Around 2013 a new tracking method was instituted named the web bug. A single pixel image file that can be hosted on any website is injected or placed on web pages. You won't see the image file or even know that it is there. This mechanism tracks the ip address of anyone viewing the page. Data brokers and the mailing list industry backed the creation of hundreds of companies collecting silently the web traffic of almost all popular websites. These bugs are still in place, software has been created to alert and block some of these trackers.

As of 2015, the problem of the interconnectedness of data is all around us at every turn. A simple click of your mouse on any webpage can create a complete picture of your identity. Not only that, a footprint is created and documented with your every keystroke. It is not commonplace for you to realize this, but you are recognized and tracked. The choices you make in the moment, in the here and now, can have lasting effects not only on your life but on those that you are associated with.

The concept of data and its originality is the underlying issue that we presently face on a daily basis. The government often unwittingly sets and creates public policy. There is a perception that creation of data is actually the real problem. Those that work as public officials and are elected to office have a bias toward the use of lists. Without lists of past donors to their respective parties, they would have difficulty raising the funds necessary to be elected and re-elected. Our information is today's crack to politicians and those running for office.

The companies that are known by consumers as the Credit Reporting Industry also actively pursue other endeavors with the data they collect. Congress merely regulates information used toward a variety of financial decisions. These companies also sell a wide variety of other reports to just about anyone for any other purposes. They are very active in insurance and health care and just about every other facet of our collective data existence.

As a society, we were coerced into the belief that we had to be part of the system. We unknowingly exposed ourselves. We then became part of a macro level of data collection that was allowed to operate without the necessary allocation of reasonable security measures surrounding the ownership of our name.

The problems we face are deep and difficult. Changing property laws to owning our own name is only part of the answer. Imagine if history dictated a different course. Would today's problems even exist if "The Right to Privacy" document written over a hundred years ago had been read instead of being cited by those writing laws and working on collage papers?

The individual shall have full protection in person and in property. Author unknown and written long before our country was a concept.

A RIGHT TO PROPERTY

Notes

1. A Robinson. Writing and Script: A Very Short Introduction. Oxford University Press, 1 Oct 2009. ISBN 0199567786.

2. Tappan (2004,June 6). Retrieved September 23, 2014, from http://www.pbs.org/wgbh/theymadeamerica/whomade/tappan_hi.html

3. BEFORE A LOAN IS OBTAINED: Precautions Taken by Banks in Investigating Would-be Borrowers.
New York Times (1857-1922); May 3, 1914;

4. Ten Things You Might Not Know About Experian - Experian Business Information Services. (2013, December 6). Retrieved September 23, 2014, from http://www.experian.com/blogs/business-credit/2013/12/06/ten-things-you-might-not-know-about-experian/

5. Article: First, Do No Harm: We Must Remove, Replace And .., Retrieved November 15, 2014 from http://www.opednews.com/articles/First-Do-No-Harm-We-Must-by-Jane-Schiff-111021-774.html_br

6. History of Credit Bureaus: Equifax, Experian, TransUnion & Innovis , Wang, Jim (2005, July ,15)Retrieved September 23, 2014 from http://www.bargaineering.com/articles/history-of-credit-bureaus-equifax-experian-transunion-innovis.html

7. How Computers Changed the Tax Game by Richard Green on April 15, 2014 Retrieved October 3, 2014 The National Archives http://blogs.archives.gov/unwritten-record/2014/04/15/how-computers-changed-the-tax-game/

8. Experian history Retrieved September 30,2014 http://www.experianplc.com/about-experian/history.aspx

9. Credit for Consumers
By LOUIS RICH New York Times ; May 10, 1931;

10. Equifax, Inc. – FREE Equifax, Inc. information .., Retrieved November 15, 2014 from http://www.encyclopedia.com/doc/1G2-2841000017.html_br

11. International Directory of Company Histories, Vol. 28. St. James Press, 1999. Retrieved November 15, 2014 from http://www.fundinguniverse.com/company-histories/equifax-inc-history/

12. Real Estate Weekly » Blog Archive » ARC building .., Retrieved November 15, 2014 from http://www.rew-online.com/2013/11/13/arc-building-portfolio-of-the-century/_br

13. Charles E. Thompson (American businessman) - Britannica Online Encyclopedia." Britannica.com. Retrieved September 25, 2014 http://www.britannica.com/EBchecked/topic/850329/Charles-E-Thompson

14. GlobalSecurity.org TRW Retrieved September 25, 2014 http://www.globalsecurity.org/military/industry/trw.htm

15. Dr. Leslie A. Hromas , The Legacy of TRW and Space Park October 2005 Retrieved September 25, 2014 from http://tra-spacepark.org/docs/TRW_History.pdf

16. Reel to Reel Tape Recorder Manufacturers - Bell Sound Retrieved November 15, 2014 from http://museumofmagneticsoundrecording.org/ManufacturersBell-Labs.html_br.

17. FundingUniverse Retrieved November 15, 2014 from http://www.fundinguniverse.com/company-histories/TRW-Inc-Company-History.html

18. Biographical Dictionary of American Business Leaders, by John N. Ingham, Greenwood Publishing Group ISBN-13: 978-0824778491

19. 19 CREDIT AGENCIES DUE FOR INQUIRY: Senators Call Witnesses to Monopoly Unit Hearing
New York Times Dec 9, 1968;

20. Sen Ervin vs. 'Information Power' Wall Street Journal, February 8, 1971

21. Computer use of credit Data called unfeasible at hearing. Associated Press ,New York Times May 17,1968

22. TransUnion, Company History Retrieved October 6, 2014 from http://www.transunion.com/corporate/about-transunion/who-we-are/company-history.page

23. Ward Body Works - FranaWiki - UCA Honors College, Retrieved November 15, 2014 from http://honors.uca.edu/wiki/index.php/Ward_School_Bus_Manufacturing_br .

24. Demographics wiki University of Arkansas Retrieved October 6, 2014 http://honors.uca.edu/wiki/index.php/Demographics,_Inc.

25. UNFAIRNESS LAID TO CREDIT AGENCY: F.T.C. Charges Deception in Reporting Tactics Company Surprised 114 Credit Bureaus New York Times; Dec 19, 1973; pg. 91

26. Computer use of credit Data called unfeasible at hearing. Associated Press ,New York Times May 17,1968

27. Credit Investigators Say They Had to Falsify Data: Purpose of Tactic Neighborhood Nut' Special to The New York Times New York Times (1923-Current file); Feb 6, 1974;

28. New Hampshire Drive for Mills Opens By BILL KOVACHSpecial to The New York Times. New York Times [New York, N.Y] 03 Feb 1972: 20.

29. Federal Trade Commission Retrieved September 30, 2014 http://www.consumer.ftc.gov/articles/0219-disputing-credit-card-charges

30. STEVEN RATTNER New Fair-Credit laws Starting Smoothly New York Times November 13, 1975

31. Demographics wiki University of Arkansas Retrieved October 6, 2014 http://honors.uca.edu/wiki/index.php/Demographics,_Inc.

32. Demographics - FranaWiki, Retrieved November 15, 2014 http://honors.uca.edu/wiki/index.php/Demographics_Inc._br .

33. Hoovers Company profiles Retrieved November 15, 2014 http://www.hoovers.com/company/Harte-Hanks_Inc/rjsfri-1-1njhxk.html

34. Hoovers Company profiles Retrieved November 15, 2014 http://www.hoovers.com/company/Harte-Hanks_Inc/rjsfri-1-1njhxk.html

35. HHS. Answers.com. Company Profile, Hoover's, Inc., 2014. Retrieved October 10, 2014 http://www.answers.com/topic/harte-hanks-inc

36. Computer Encyclopedia HHS on Answers.com. Computer Desktop Encyclopedia Copyright © 2014 by Computer Language Company Inc.. Published by Computer Language Company Inc..

37. HHS. Answers.com. International Directory of Company Histories, The Gale Group, Inc, 2006. Retrieved October 10, 2014 http://www.answers.com/topic/harte-hanks-inc, accessed

38. Wired magazine , Issue 3.09 | Sep 1995 Retrieved October 06, 2014 http://archive.wired.com/wired/archive/3.09/equifax.html

39. Information Today;May1996, Vol. 13 Issue 5, p39 Retrieved October 10, 2014 http://connection.ebscohost.com/c/articles/19074729/pro-cd-inc-acquired-by-acxiom-corporation

40. Bloomberg Business week company snapshot retrieved October 10, 2014 http://investing.businessweek.com/research/stocks/private/snapshot.asp?privcapId=214538069

41. Information Today;May1996, Vol. 13 Issue 5, p39 Retrieved October 10, 2014 http://connection.ebscohost.com/c/articles/19074729/pro-cd-inc-acquired-by-acxiom-corporation

42. Frontline United States Secrets part one PBS originally aired 5/13/2014

43. The blaze, Google Warns of Latest Online Security 'Vulnerability' Jon Street October 15, 2014 Retrieved October 16, 2014 http://www.theblaze.com/stories/2014/10/15/google-warns-of-latest-online-security-vulnerability/

44. A Brief History: The Bell System Retrieved September 26, 2014 http://www.corp.att.com/history/history3.html

45. AT&T Business Analysis - Research Paper, Retrieved November 15, 2014 http://essays24.com/Business/AtT-Business-Analysis/6392.html_br .

46. The Bell System| History| AT&T, Retrieved November 15, 2014 http://www.corp.att.com/history/history3.html_br.

47. February | 2013 | Village Book Shop, Retrieved November 15, 2014 http://villagebookshopglendora.wordpress.com/2013/02/_br .

48. Jason Zasky. "The Phone Book." Failure Magazine. Retrieved 2013-12-31 from http://failuremag.com/feature/article/the_phone_book/

49. Majority of U.S. Mobile Phone Users Against Development of Wireless Directory. PR Newswire Association LLC Dec 7, 2004 Retrieved Septermber 24, 2014

50. Yellow Pages ruling endangers SF ban, Heather Knight, San Francisco Chronicle, October 15, 2012; accessed March 19, 2013 http://www.sfgate.com/bayarea/article/Yellow-Pages-ruling-endangers-SF-ban-3951477.php

51. Badore, Margaret (2010-01-11). "Ask Pablo: What Is The Impact Of All Those Unwanted Phone Books?." TreeHugger. Retrieved 2014-04-16. http://www.treehugger.com/culture/ask-pablo-what-is-the-impact-of-all-those-unwanted-phone-books.html

52. Branscomb, Anne (1994). Who Owns Information?: From Privacy To Public Access. BasicBooks. ISBN-13: 978-0465091751

53. 5 U.S.C. § 552(a)(4)(F) http://www.law.cornell.edu/uscode/text/5/552 Retrieved September 26, 2014

54. "FOIA Legislative History." The National Security Archive. The National Security Archive. Retrieved 24 September 2013. http://www2.gwu.edu/~nsarchiv/nsa/foialeghistory/legistfoia.htm

55. Gerhard Peters and John T. Woolley. "Lyndon B. Johnson: "Statement by the President Upon Signing the "Freedom of Information Act.," July 4, 1966.." The American Presidency Project. The American Presidency Project. Retrieved 24 September 2013. http://www.presidency.ucsb.edu/ws/?pid=27700

56. Metcalfe, Daniel J. (23 May 2006). "The Presidential Executive Order on the Freedom of Information Act" (PDF). 4th International Conferene of Information Commissioners. pp. 54–74. Retrieved 20 June 2013. http://www.edps.europa.eu/EDPSWEB/webdav/site/mySite/shared/Documents/EDPS/Publications/Speeches/2006/06-05-23_ICIC_2006_Manchester_EN.pdf 57

57. A Summary of Your Rights Under the Fair Credit Reporting Act Retrieved September 23, 2014, from Federal Trade Commission, https://www.consumer.ftc.gov/articles/pdf-0096-fair-credit-reporting-act.pdf

58. EPIC - The Fair Credit Reporting Act (FCRA) and the Privacy Retrieved November 15, 2014 http://epic.org/privacy/fcra/_br .

59. Sale of Financial Data Starts a Backlash; Privacy: California is expected to be in the fore of a push by consumer groups and lawmakers to give bank customers more control over the sale of information about them Sanders, Edmund. Los Angeles Times [Los Angeles, Calif] 12 Nov 1999: A, 1:6.

60. How To Comply with the Privacy of Consumer Financial Information Rule of the Gramm-Leach-Bliley Act Retrieved September 23, 2014, from http://www.business.ftc.gov/documents/bus67-how-comply-privacy-consumer-financial-information-ruule-grammleach-bliley-act

61. Facebook friends could change your credit score CNN MONEY By Katie Lobosco @KatieLobosco August 27, 2013: 11:24 AM ET Retrieved September 26,2014 http://money.cnn.com/2013/08/26/technology/social/facebook-credit-score/

62. Erin Lowry. "How Bad Moves on Social Media Could Damage Your Credit Score - DailyFinance." DailyFinance.com. DailyFinance.com, n.d. Web. 26 Sept. 2014.

63. MilleniumBank.gr - Banking in the new Millennium, Retrieved November 15, 2014 http://www.millenniumbank.gr/MIllenniumVB/el-gr/Branch.aspx?Type=Branch_br .

64. JEREMY QUITTNER Banks to Use Social Media Data For Loans And Pricing AN 6, 2012 9:47am American Banker Retrieved November 15, 2014 http://www.americanbanker.com/issues/177_18/movenbank-social-media-lending-decisions-brett-king-1046083-1.html

65. Mining Social Media - CUNA Technology Council, Retrieved November 15, 2014 http://www.cunatechnologycouncil.org/news/4715.html_br

66. David, Goldman (Feb 23, 2012). "White House pushes online privacy bill of rights." CNN Money. Retrieved October 16,2014 http://money.cnn.com/2012/02/23/technology/privacy_bill_of_rights/index.htm?iid=EL

67. The White House, "Consumer Privacy Data in a Networked World" (Feb 23, 2012), Retrieved October 16,2014 http://www.whitehouse.gov/sites/default/files/privacy-final.pdf

68. Jennifer, Valentino-DeVries (Mar 29, 2012). "Yahoo to Implement 'Do Not Track' Mechanism." The Wall Street Journal. Retrieved October 16,2014 http://www.youtube.com/watch?v=HutfN-Pu0Mc

69. Rainey, Reitman (Jan 24, 2011). "Mozilla Leads the Way on Do Not Track." Electronic Frontier Foundation. Retrieved November 15, 2014 https://www.eff.org/deeplinks/2011/01/mozilla-leads-the-way-on-do-not-track

Privatize yourself: Taking back control

"All compromise is based on give and take, but there can be no give and take on fundamentals. Any compromise on mere fundamentals is a surrender. For it is all give and no take."
Mahatma Gandhi

The necessity of taking control of your information is crucial not only to yourself but your family and friends. Several steps are needed. The first step begins with seizing your name and disrupting the flow of information to the brokers of your information. It will also include some forms of pay back and some fun if you're up to it. Data brokers are the current bogie man in many newspapers. Some of those in the government want to legitimize the activity through regulation. This step would forever make it legal for companies to monitor our actions.

There are many flaws in the system you should be aware of. The companies that collect information are very vulnerable to their own greed. Many of the systems in place were spawn from the mailing list industry. The data collections are for the most part setup in lists of categories. The data collection system is primarily additive and blindly adds information to your profile without any checks or validity. It is not anticipated that a large number of consumers would buy products that he or she does not have any interest in and then immediately return them. This is only one flaw, there are others.

There is not currently a uniform system that verifies data. Understanding how the lists works is very powerful. Knowing how it functions is crucial because it allows the owner of the information to make it invalid. Those that conduct marketing research expect that the data that is collected will suggest who in the target demographic will purchase the goods.

By virtue of joining lists of things you are not interested in, you are messing up the system. It may seem insignificant with the information being held on at least 600 million people. The population of the United States is just over 318 million people. If the marketing databases showed a dramatic discrepancy from national statistics, then the list would be useless. There are many lists that house our information and are used against us on a daily basis. These lists conjoin to create a form of marketing intelligence. The idea is to have a running picture and or a psychological profile on just over half of the U.S. population. For further information, take a look at the website http://itsmyinfo.org for the monthly list target.

It is doubtful that sending a single letter to the corporate headquarters of your local drug store chain will change any of the existing policies that have been put in place. If the same chain has a thousand people demand that a change in policy occur then it is more likely that the company would need to respond.

Step One: Notification

The first step in taking action begins with the official notification of intention. The United States Post Office offers a certified mail and a return receipt service. This is the preferred method and legally acceptable form of communication that the courts recognize. It commands authority and is proof that someone sent something to another party. It is my recommendation that you always make copies of anything that is mailed. Make a folder and save copies of paper correspondence that includes receipts from the post office that are stapled or taped to your photocopies. This form of record keeping may prove useful in step three.

All of us have relationships with physical companies and internet websites. The employees of these companies are in the same boat as you. As individuals in our society, we have been romanced into a false belief that our information is not ours. Many employers that receive your correspondence are guilty of selling our information. What the company does is not directly the fault of the employee and keep this in mind when you reach out to the establishment. Enlighten the companies and the employees and explain your actions. Tell them about this book and what you have learned.

The following example letter should be checked by your own legal counsel before mailing it to everyone that you do business with. You can download this form directly from http://itsmyinfo.org

Today's date

Personal and Confidential

Attention legal or compliancy department.

Company name

Address

City State, Zip

Your name

Address

City State, Zip

Addendum to Privacy policy/statement/notice of Company name

I have an active business relationship with your organization and am giving advanced notice that selling my information is no longer permitted. Your right to use my name and other identifiable information is hereby restricted to transactions dealing only with myself. You are not permitted to sell or share any identifiable information with any third party except governmental agencies without direct, advanced and timely notification and individual approval by me. For your consideration, I have enclosed an individual third party release of information form. Each and every vendor that you may disclose my information to requires a release of information form signed by myself.

Feel free to copy the third party release of information form as many times as necessary.

Thank you in advance for your consideration.

_____

_____   _____

                                    Signed

          Dated

Enclosures: 1

Sample Third party release form.
Company name
Address
City State, Zip
Your name
Address
City State, Zip

Our company must transact information to _____
_____company name_____
_____Full   Address   and   contact
information_____
    In the course of doing business, please be advised
that we disclose identifiable information for the following
reason.

_____

_____

_____

_____

[ ] I hereby allow the use of my name to be loaned to
_____ Company name_____ for a limit of
_____ days or years.

Signed and dated by _____

_____

- -

- -

Take the information from the Third party release form
mailed back to you and send the first form to the address
listed. Most people only deal with a limited number of
companies. Start slow with your utilities, grocery store,
and the banks you do business with. The companies on the
internet that house our information are a bit trickier than
the public utilities. Many of them hide their address and
contact information. Owners of any website can be found.
Post any questions that you may have to the forum on website
http://itsmyinfo.org/fourm

A number of the collectors of information that are hidden.
Compelling their identities will require legal action. Those
that do not have any user agreement signed and on file will
be the first targets of civil remedy. It is likely that attorneys
will need to be enlisted in this endeavor. Lawsuits and courts
action will be a long process. The action of change will not
happen quickly.

Step two: Invalidate your information

The entire industry depends on your information being sort of correct. Being near correct is what makes it valid. There are things that you can do that cost nothing and can change the information about you. One of the most powerful statements you can make is to purchase items with a loyalty card and pay with a credit card that you don't like or want. If you want to buy can goods for your local food drive consider buying things you are allergic to or do not like. If you are going to purchases items and return them, remember the rules of the store you are dealing with. Also don't buy and return things more than once a month to the same store or chain.

Try to search for thing on the internet that you are not interested in. Many of the search engines list sponsored ads at the top. These are advertisers that pay the search engine to display their advertisement. By clicking on these will also add items of interest to your profile. If you use any of the social networks you will see new pop-up advertisements. It can become your own private joke.

There are applications that you can put on your phone to spoof your GPS position. I personally like to set mine at home whenever I don't need it. According to google I hang out at home a lot. Tracking still can be done from cell tower data. The best advice is do not put applications on your phone unless you understand where the data goes. I have tried a few apps to block access on my phone. There is not one that I would recommend at this point.

Step three: legal remedies

Sharing is not the same as selling and is done without any form of compensation. Almost all of the agreements you have signed specify that your information may be shared. Any organization, store, bank or website that you have joined has a policy that you may have agreed to. Your agreement was given to you as a take-it-or-leave-it arrangement in order to do business or obtain a discount. It is likely that your agreement may be in breach if any form of compensation has taken place.

Suing for compensation of selling information will take many plaintiffs. That is simply because even though it is your information, a single store may only receive pennies per transaction. No court would allow suing for a few dollars. If a few thousand people all shared a claim for a couple of dollars each, it would be more likely that the case would be heard.

The idea that your name has any value as property really doesn't have a foundation within court case history. The reason behind that is that the American people have been taken advantage of so completely that the very perception within general society is overpowering. There is nothing that anyone can do. This can be a tactical advantage because the foundation for decisions can be used later to shape smaller state case rulings.

There are areas in the country that a civil suit would have a better chance of receiving favorable rulings. There will be others that will be sympathetic to the idea that your property has been violated. The problem is so engrained that judges and members of the court are victims too. They will need to be sought out and identified. Many attorneys jump at the chance to represent class action lawsuits. That type of law is not suitable and has many drawbacks for the population. Attorneys like it simply because they keep most of the money from the case. As a result, clients receive very little compensation.

Another problem with class actions is that even if a case is won, the company can only be sued once. Meaning that the company can go right back to what they were doing without needing to worry about being involved in further litigation. The tort law system is broken and really needs a book written to address this matter. Perhaps, the title 'Pennies On the Dollar, The Truth About Tort Law" or "Economic Leech's in American Society"

Starting at the top is foolish because big business equates to a massive war-chest. Large companies can afford expensive and more experienced law firms that could possibly stop us in our tracks. Smaller companies do not have enormous budgets and cannot afford expensive law firms. Defendants in litigation can ask for a change of venue, which would potentially be held in federal court.

Websites have headquarters in various states. Some are just in the business of data collection and appear as something entirely different. An example of this is a company in Florida that has a message board for grandparents. It is a facade for people openly to post messages about problems they have with their grandchildren.

Other topics include investing in their grandchildren's future. Users on the website are unaware that it and many like it have hidden agendas. The sites are for the sole purpose of getting personal details from people. This specific company is not a small enterprise. There are many companies on the internet that can appear as small enterprises, but are actually part of larger corporations.

It would be wiser to reserve court actions to smaller operators to start with. It will take some time to build a complete legal strategy. It also takes talented, interested and legal minds to collaborate with on the project. If you are an attorney, please register on the website provided below, and you will be contacted. http://itsmyinfo.org/attorney

It would be wise to use the courts in concert with media releases and protests. This will gain media attention to the matter. Television can be a powerful ally or an enemy of any movement. Social networking sites such as Google, Facebook, Meetup, and Twitter are ideal for spreading the word. I encourage you to tell others about this book and get our personal information off the internet. We did not place it there ourselves. We should not be forced either to have our personal information shown on the internet.

There are very big and powerful corporations that will be opposed to changing the selling of everything we do. There is a looming possibility that those with big money will learn about this writing. Those with deep pockets will create advertisements and enlist the government to pass laws to protect the information that has been stolen from us. As of this writing, data brokers are attempting to be legitimized by Congress. The legal system and those in Congress have the ability to rectify much of the data leakage that occurs today. Information about the general population should not be allowed to be made public. Our own names are personal property, and we should be legally allowed to keep our information kept as such. We do not have to behave as virtual slaves simply because the government has our names.

Step Four: contact federal legislators

One of the first things that legislators should do is revise the HIPPA statutes. Use abstract language that states that all products humans use externally or internally outside of clothing fall under HIPPA. Data brokers sell information; credit companies collect who is ill. They assume though what was purchased paying for medication using loyalty or credit cards puts you on lists. Aspirin, toe nail clippers, tampons, and everything else should be items that so one has the right to know about the sale of. The legislation should force companies that collect information and sell it to compensate the victims of data collection. We should be the only ones who own our data and when our information exchanges hands, we should be allowed to provide the permission to do so. As the owner, we deserve the right to be fairly and justly compensated for the data collected associated with not only our name but our daily interactions.

The Federal government already has an agency in place that is for watching out for the protection of consumers.

This was taken from the website in January of 2015.

"In June 2009, President Obama proposed to address failures of consumer protection by establishing a new financial agency to focus directly on consumers, rather than on bank safety and soundness or on monetary policy. This new agency would heighten government accountability by consolidating in one place responsibilities that had been scattered across government. The agency would also have responsibility for supervision and enforcement with respect to the laws over providers of consumer financial products and services that escaped regular Federal oversight. This agency would protect families from unfair, deceptive, and abusive financial practices. The President urged Congress to give the consumer agency the same accountability and independence that the other banking agencies have and sufficient funding so it could ensure that powerful financial companies would comply with consumer laws.

In July 2010, Congress passed and President Obama signed the Dodd-Frank Wall Street Reform and Consumer Protection Act. The Act created the Consumer Financial Protection Bureau (CFPB). The CFPB consolidates most Federal consumer financial protection authority in one place. The consumer bureau is focused on one goal: watching out for American consumers in the market for consumer financial products and services."

The companies that sell our information all claim that they are doing so under the guise of marketing information. During a 60 Minutes episode that aired on August 24, 2014 entitled "The Data Brokers," Correspondent Steve Kroft narrated a story that examined some of the issues related to data brokers. During the program, Brian Kennedy, the Chairman and CEO of Epsilon, was interviewed. The company was represented in the report as having the world's largest corporative database. The data includes more than "8 billion consumer transactions with an extensive network of online sources." During the interview, the claim was made that data collection is the fastest growing sector of the economy. Mr. Kennedy stated during the interview that regulation would potentially "cripple the economy."

By the industry's own admissions before Congress, the issue is a financial one centered around consumers. The Consumer Financial Protection Bureau should be put in charge of watching over the information collected about consumers and work with the states and make the practice of wholesale selling of identifiable information an illegal practice.

Not only that, but we should be informed of who and what companies have access to our information at all times and be informed of where it is housed and when it is disseminated. A chain of custody needs to exist when our personal information is used. Companies do not need to keep information about you as the consumer longer than a year unless given permission in the form of a customer-bought warranty. It should become a criminal act to hoard private information that identifies individuals.

This very matter begs you to re-visit credit reform. Force credit reporting companies to agree that anyone has the right to monitor their credit anytime they want. Individuals should have access as many times as they want in any given calendar year. Create legislation that fines the credit reporting industry for failing to adhere to the legal obligation of legislation. Fines should be imposed for holding information that is outside of the length of disclosure. It is my recommendation that you further restrict the information to only financial transactions. Credit companies should not be allowed to use financial information in any other fashion. It should be illegal for any data to be used for any other purpose. There should be a firewall put in place for credit reporting companies that would bar them from engaging in any other business. They should be forced by regulation to be split apart and act as entirely separate organizations from the other business they currently operate alongside. Credit reporting organizations are the original and biggest data brokers that actively harvest and sell many pieces of information though their other lines of business The only reason these companies collect information is to sell it. All of the statements that these organizations make about security or our own economy are just smokescreens for protecting a wide variety of other intrusive actions.

The Committees of Congress should add the Homeland Security license reading program to the NCIC system. The license information should be closed and be accessed at the discretion of the Attorney General and the American people. Private industry should pay for its own workforce. The act of using law enforcement agencies and local communities to collect information for any private company to freely market is disgusting.

It is reprehensible that our local police are spying on our movements. The entire system should be abandoned, but I understand why it is present.

We the people need the government to do its part.

The following list of changes to U.S. Code covers some of the topics covered in this book. The repercussions of change are daunting, even to the legislators that need to be contacted.

Lobby your federal legislator to alter the following:

Title 18 › Part I › Chapter 103 › § 2112 to read Personal property.

Whoever robs or attempts to rob another of any kind or description of personal property ~~belonging to the United States~~, shall be imprisoned not more than fifteen years per occurrence.

This makes it a federal crime for anyone that steals property of any kind throughout the nation.

Alter Title 15 › Chapter 94 › Subchapter I › § 6801 to protect all institutional data (not just financial institutions)

(a) Privacy obligation policy

It is the policy of the Congress that each ~~financial~~ institution has an affirmative and continuing obligation to respect the privacy of its customers and to protect the security and confidentiality of those customers' nonpublic personal information.

(b) Financial institutions safeguards

In furtherance of the policy in subsection (a) of this section, each agency or authority described in section 6805

(a) of this title, other than the Bureau of Consumer Financial Protection, shall establish appropriate standards for the financial institutions subject to their jurisdiction relating to administrative, technical, and physical safeguards—

(1) to insure the security and confidentiality of customer records and information;

(2) to protect against any anticipated threats or hazards to the security or integrity of such records; and

(3) to protect against unauthorized access to or use of such records or information which could result in substantial harm or inconvenience to any customer.

(4) to notify each customer when information is lost. Require the Attorney General to investigate and prosecute those that fail to adhere to standards of reasonable safeguards of information.

(c) institutional safeguards

(1) to insure the security and confidentiality of customer records and information;

(2) to protect against any anticipated threats or hazards to the security or integrity of such records; and

(3) to protect against unauthorized access to or use of such records or information which could result in substantial harm or inconvenience to any customer.

(4) to notify each customer when information is lost. Require the Attorney General to investigate and prosecute those that fail to adhere to standards of reasonable safeguards of information

Strike all references to publicly available information as applies to individuals throughout U.S. Code.

Alter 15 U.S. Code § 6809 - Definitions
(C) Notwithstanding subparagraph (B), such term—
  (i) shall include any list, description, or other grouping of consumers (~~and publicly available information pertaining to them~~) that is derived using any nonpublic personal information other than publicly available information; but
  (ii) shall not include any list, description, or other grouping of consumers (~~and publicly available information pertaining to them~~) that is derived without using any nonpublic personal information.

Additionally, lobbing your federal legislators to create a right to property law is something that could be done. As the book spells out it is your property anyway. It is unlikely that legislators will listen to us. They are hooked on the lists as much as the corporations that fund their re-election efforts. It could be entered in a single line in the following statue:

Alter Title 42 › Chapter 21 › Subchapter I › § 1982
a)All citizens of the United States shall have the same right, in every State and Territory, ~~as is enjoyed by white citizens~~ thereof to inherit, purchase, lease, sell, hold, and convey real and personal property.
b) All citizens of the United States shall have the right of property in their name and identity.

Public access to the National crime database would solve many of the inaccuracies in that system. It would also shut down the various inaccurate data bases that are present on the internet.

Alter Title 42 › Chapter 151 › Subchapter I › Part C › § 16961
(a) In general
Notwithstanding any other provision of law, the Attorney General shall ensure access to the national crime information databases (as defined in section 534 of title 28) by—

(1) the National Center for Missing and Exploited Children, to be used only within the scope of the Center's duties and responsibilities under Federal law to assist or support law enforcement agencies in administration of criminal justice functions; and

(2) governmental social service agencies with child protection responsibilities, to be used by such agencies only in investigating or responding to reports of child abuse, neglect, or exploitation.

(3) the general public

(b) Conditions of access
The access provided under this section and associated rules of dissemination, shall be—

(1) defined by the Attorney General; and

(2) limited to personnel of the Center or such agencies that have met all requirements set by the Attorney General, including training, certification, and background screening.

(3) open to the general public at the discretion of the Attorney General

Step Five: contact your state legislators

Your state is in the data selling business. Your tax dollars are supporting a multibillion dollar a year business and it does not need your money. Ask your legislators to audit what data is leaving the state and for what purpose. Software used in the course of government business should not automatically send information to 3rd parties. It should be the citizen's choice if they want to participate. If brokers want to pay each and every individual tax in exchange then, perhaps some people would be interested. I am not interested; it should be an individual's choice. State programs that give personal identifiable information to brokers should be abolished. Call your department of motor vehicles and ask who has data outside of government. Ask your legislators to cut the cord we do not need our states selling our personal information.

Step Six: protect your own information

Personal security starts at home. Purchase a cross-cut shredder if you don't have one. Shred any paper that has your name on it. This item is very inexpensive and can save you from future heartache from those who want to capture and utilize your information.

There are various keys of your identity which include your name, address, date of birth, and social security number. Your identity needs to be guarded; you can start by paying attention to whom and for what reason you are providing your personal information. Never enter real personal information on websites. The internet has never been secure, and it is doubtful that it ever will be.

Security questions are a constant problem for a number of reasons. These very questions provide a false sense of security to society that our information is secure. Unfortunately, there are companies that create questions and queries to be used by major corporations to uncover and house personal details of their customers. Companies that collect information for selling it also harvest security questions to use in your profile.

As a result, many websites have been compromised, and the answers have been taken. Many individuals also post answers to their security questions on Facebook and other social media websites. Schools and dates of birth are things that identity thieves look for. Many sites allow changing the information. No site on the internet needs to know any of your keys to identity.

The social security number is especially vulnerable and valuable. Many companies insist on using the last four digits as a security question. This question alone allows the consumer to become a victim. The first three digits are the state where the card was issued. This leaves a two digit number that is called a group number that is also from the state where the card was issued. Two people that register for numbers in the same year and from the same state around the same time will have a virtually identical first five digits.

The last four digits of the Social Security Number are the most important ones to protect. I have called and asked companies I deal with and ask to change the number on file to another number made up.

It is not my suggestion to contact companies you deal with that use this type of security and change the social security number they have on file. Just write the number you use down, so you don't forget it.

All of your security questions and password should be different. Write them down on a sheet of paper do not save them on the internet. One approach is using movie stars or sport figures for all of the answers. Just don't answer any question using your own information and never use the same answer twice.

I have a free paper on my website to avoid identity theft.
http://tripelix.com

You no longer need to be fooled or be under the guise that you are afforded some level of protection once you provide your information to a company. The efforts of even small organizations to obtain personal information come in various ways. The best answer to their collection efforts is to tell them no. Be tactful and polite; don't give information they don't need. A grocery store wanted my date of birth for a loyalty card. I made up a date I have no idea what it is. Companies simply do not need information they ask for. Ask yourself, "why do they need that" anytime you are asked for any personal information. Don't just blindly hand the keys of your identity to anyone.

Step Seven: Meet with others

Letter writing and legislative action and making lawsuits are a way of starting progress. As your author, I encourage you to take action. This cannot be done alone, and it is our information that is at risk. It should be easy for you to find anyone else who will listen. It can become a struggle at times to create action, but it is worth it. An individual can arm him or herself with the proper safeguards, knowledge and tools to become a viable consumer in the marketplace. Organizing yourself in dealing with your personal information is key.

Why organize?

As a modern-day society, we are faced with numerous issues associated with our identity. There are many companies that sell our information. Government officials and legislators have ignored the call of the people for too long. Congress as of this writing is attempting to legitimatize data brokers as a real industry. The entire industry has stolen our information and sells it in the open market. There will be many forces that will oppose the change that all of us currently face. A change requires spreading the message that our names should be treated as our own personal property and not afforded to governmental establishments or private corporations.

Meeting with others to enact change is imperative. Please spread the word about the ideas that you have read in this book and take action. Creating a movement will not happen overnight. If you are not a self-starter, or if you are not the type to start a movement, read along; you might change your mind.

Effective organizing requires an investment on your part. It requires that you motivate others to take charge. Learning how to organize is easy. The world is ready and waiting for you; your friends will be the foundation that you will build on.

True leadership fosters other leaders who will recruit their own teams. You must forge new relationships in the community and center your commitment around that leadership. The community will respond with resources—someone who can print signs or makes videos, embrace them. Let other leaders know about them too.

One of the primary jobs in being an organizer is to identify and recruit volunteers to work with you to build campaigns. Each person is different; we all have individual strengths and weaknesses. It is your responsibility to identify what the values are that others possess.

Don't try to be the center of attention; foster leadership with others. It will be a shared effort. Many others will have difficulties and if you need help visit the forum.

http://itsmyinfo.org/fourm Post questions there.

Ultimately, you need to decide who is responsible for coordinating everyone and who will be able to make decisions for the group when you're not available.

The shared goal of changing the attitude of millions of people is daunting. Work within your own community, create or go to meetings and meet others face to face. A true leader recognizes that delegation of responsibility and empowering others to take their own responsibility is ultimately more powerful than getting others to follow you though an individual task.

You will never have enough people around your efforts. Many people will not follow through; it is normal that people will misrepresent their abilities and will not show up for events. We all have lives and priorities within them. No matter how passionately you speak and depend on others, they will disappoint you. It is a hard lesson to learn, but it is a valuable lesson. Sometimes people will do only part of what they say they will do. Expect it so you won't be surprised. It is important that you understand who will come through, who is committed, and who can lead on their own.

It is important to work with other leaders, followers to any movement need leadership. If a group loses its leader, you will need a way to contact the group and take over leadership. Keep this in your mind and also foster this idea to others that you encourage to lead. The person leaving the group could be you. Think of other leaders as your backup and you as theirs.

Encourage others to tell their own stories. Pay attention and listen to them and let them bring alive their own passions. Let them know that they are not alone. There are things that they can do in their lives to help change the prevailing attitude that everyone's names are up for grabs. The topics can be anything: identity theft, strange marketing phone calls, strange magazines mailed to your home, and the loss of your credit card information are all great topics. Most of today's problems are caused by the forces that harvest our information.

If you know a business owner, ask them to think about whose information they have. Encourage them to adopt a voluntary policy of destroying rather than keeping personal identifiable information about customers that is older than a year. Give them a copy of this book. There are bulk discounts available. Check my website http://tripelix.com

Be prepared; there are many people who will listen to you. The like-minded will be onboard and many will ask what they can do to help. Have materials available for them to read. You can download and print forms from the website http://itsmyinfo.org/; feel free to make copies. There are a few documents there to download and print at home.

Tell them about the book or make buttons, wristbands, or shirts; here are a few slogans: "Quit being for sale," "None of your business," "I belong to me," "My name My property." Maybe you have a designer friend that can help they, can sell the shirts and keep the money.

Build Your Teams

Why does organizing teams matter?

Teams create motivation for each other. Groups of people support each other in joint efforts. It is far easier to accomplish tasks by enlisting others to participate. They can also bring different skill sets and viewpoints. Groups make experiences more entertaining and fun. They make it possible to achieve more than any individual could ever create alone.

Teams also create a hierarchy of responsibility. Members often will know what each other's strengths are. Building clear thought-out strategies and areas of responsibility are important. There have been too many failed and polarized movements in the past to list. In recent history, I would like to point out two.

Use the Occupy Wall Street movement as an example. It is to blame for many of the new laws restricting assembly. It did not have a structure or have real leaders for the media to interview. The overall message was rapidly lost. At first the protest was embraced by the media. Money and influence were used to criticize the protesters. Propaganda was created by Wall Street bankers, and public perception of the movement went away. The group was classified as nut cases; drugs and orgies were reported on the nightly news. The goals of the protesters were lost. They did not embrace the central core of Americans.

Another example is the Tea Party movement. Its primary mission was to lower taxes; however; it got hijacked by other special interests and because of it being embraced by the conservative movement, it was labeled as being racist. That might sound silly, but that is how it is portrayed in many media stories. Now it is polarized as only conservative ideas and philosophy and away from just spending issues.

Both groups would go farther if they adopted philosophy of a separation of corp and state. Corporate interests are too engrained within our government and really need to be separated. Visit our national museums and see the advertising used by our defense industry. The separation of church and state has created a vacuum effect filled by money and corporate interest. The undue influence on our legislators could fill several books that I will not be writing.

It is of utmost importance to not allow polarization of either party politics to invade the movement. Currently, there are no political views or religious beliefs that make it justified for companies to steal your information and sell it. Politicians are dependent on data brokers for their own election campaigns. Very simply, the forces that should be watching over us are actively purchasing our information. In comparison with the war on drugs, they would represent the true junkies, but with a greater addiction—namely power.

All efforts need to be in concert to be effective. Letter writing, lawsuits, and public awareness are crucial. Since the targets in this endeavor are large and have political favor, it will be an uphill battle from the start.

Having objects that represent the movement and its ideas are important. Symbolism is crucial. Use the example of the ribbon campaigns or the yellow bracelets that represent cancer research and charity. I do not have a solution here; I do not want to dictate my preference either. I will leave it to you the leader to find the symbolism and stop our information from being stolen.

Protest where data stealing is embraced. Throughout the country are "big data" conventions. These meetings are usually about how to organize the data that has been stolen from all of us. Signs and banners against this practice with media reporting makes a point. Asking for your name to be a right of property is a message.

It would be helpful for a campaign to be built around the movement. The campaign could include something simple like a sticker to place at point of sale machines. The sticker could say "no information sold;" doing this at enough locations would raise interest of others in public. Eventually this would stop the information gathering at the credit side of things. Many people would not go into a store and buy something if they had a choice not to be tracked.

It's going to be a long road; stay in touch.

Your author,
Trip Elix

Trip Elix is a published author, investigator and privacy advocate. His articles have appeared in daily and weekly publications in the United States. He writes both fiction and non fiction. He also writes a blog that focuses on security and privacy issues at http://tripelix.com.

In the past he has been a forensic computer technician who has used or owned almost every version of every operating system used since the 1980's. He regularly attends security conferences held by the computer underground, including Defcon , bsides and others.

If you want to know more about author, Trip Elix and use social media he maintains a presence on most of them.

twitter @trip_elix
https://www.facebook.com/tripelix
https://www.linkedin.com/pub/trip-elix/89/548/673/
https://plus.google.com/u/0/+TripElixauthor

www.ingramcontent.com/pod-product-compliance
Lightning Source LLC
Chambersburg PA
CBHW060444240326
41598CB00087B/3427